Food Microbiology
Processing Technology and Feed Additives

ABOUT THE AUTHOR

Dr. Subha Ganguly, B.V.Sc. & A.H. (Gold Medalist), M.V.Sc. (First Rank), NET Qualified, Ph.D. (Microbiology), Executive-MBA (Human Resources Management), D.Sc (*Honoris Causa*) is serving as Scientist (Food Microbiology) in the All India Coordinated Research Project on Post Harvest Technology (ICAR) at Faculty of Fishery Sciences, West Bengal University of Animal and Fishery Sciences, Kolkata, WB, India. Dr. Ganguly is Fellow, Society for Applied Biotechnology, India [FSAB], Fellow, Hind Agri-Horticultural Society, India [FHAS], Fellow, International Science Congress Association, India [FISCA], Fellow, Institute of Integrative Omics and Applied Biotechnology, India [FIIOAB], Fellow, International Society of BioTechnology, India [FISBT], Fellow, Academy for Environment and Life Sciences, India [FAELS], Fellow, Society for Advancement of Sciences, India [FSASc], Fellow, Society of Education, India [FSOE], Fellow, Science and Education Development Institute, Nigeria [FSEDInst], Fellow, Association of Pharmacy Professionals, India [FAPP], Fellow, Vital Biotech Research & Training Institute, India [FVBRTI], Fellow, Pharmacy & Life Sciences, India [FPLS], Honorary and Executive Board Member, Pharma Research Library, India (PRL), Executive Committee Member, Research Scholar Hub, India (RSH) and Society of Researchers and Health Care Professionals, India (SRHCP) and an honorary eminent scientist by the International Biographical Centre, Cambridge, England and the American Biographical Institute, USA. Dr. Ganguly has been conferred with the distinction "American Order of Merit" from the American Biographical Institute, USA. Dr. Ganguly is also a prominent member of many reputed National and International scientific societies and awarded with the distinctions MNAVS- Associate Member/ Fellow NAVS India, MISZS, LMIVA, MVPHA, MIPHA, LMJZS and MAPHV. Dr. Ganguly has many high quality peer reviewed publications in highly reputed and indexed National and International journals and he has also authored many books and monographs on specialized issues of concern. Dr. Ganguly is in the capacity of Editor-in-Chief, Associate Editor, Member Editor, and Honorary member of editorial boards and advisory committees of numerous National and International indexed journals of repute with many honors, awards and distinctions from various scientific bodies. Dr. Ganguly has authored numerous books on prominent and specialized topics in the field of Veterinary & Animal Sciences, Fishery Sciences and Food Sciences for the expansion of knowledge among interested Students and Professionals.

Food Microbiology
Processing Technology and Feed Additives

Dr. Subha Ganguly

2015

Daya Publishing House
A Division of
Astral International Pvt. Ltd
New Delhi-11002

Published by	:	**Daya Publishing House®**
		A Division of
		Astral International Pvt. Ltd.
		– ISO 9001:2008 Certified Company –
		House No. 96, Gali No. 6,
		Block-C, 30ft Road, Tomar Colony, Burari
		New Delhi-110 084
		E-mail: info@astralint.com
		Website: www.astralint.com
Sales Office	:	4760-61/23, Ansari Road, Darya Ganj
		New Delhi-110 002 Ph. 011-23245578, 23244987
Laser Typesetting	:	**GRB 7color Service**
		New Delhi - 110 084
Printed at	:	**Replika Press Pvt. Ltd.**

PRINTED IN INDIA

Preface

The book provides the advance and updated information on various issues related to hygiene and quality control in food processing sector. The book encompasses on the broad aspects of immense concern in the field of Food Microbiology and Processing Technology. Even it finds its broad implication in processing industries and breweries. The book can be used as textbook and a ready guide by the Students of food Processing Technology and allied sciences, as well as the professionals associated with the ever expanding food processing sector.

The book is dedicated towards the development of Science, Education, Knowledge and Technology.

Dr. Subha Ganguly

Contents

Chapter 1
Introduction

Post-harvest losses of vegetables, fruits and fisheries are difficult to predict, the major agents producing deterioration mostly being attributed to microbiological causes and physiological damages. Post-harvest losses may be grouped broadly into food losses after harvesting and food losses due to social and economic reasons.

The losses at each stage of harvest and post-harvest practices due to improper handling can be large enough to result in a total loss of millions of food commodities every year. It is believed that a 50% reduction in post-harvest food loss in developing countries will reduce the need for food importation in these countries and will cause an increase in the food supply to meet the food demands. Also loss is far less than the amount of money that will be used to produce the same amount food.

However, it has been discovered that increased food production only is not the final solution if it is not complemented with adequate harvest and post-harvest practices. This is because good harvest and post-harvest practices will lead to reduction in the amount of food losses during and after harvest. The food moved from the farm, through the delivery system to the consumer must be presented in the good and acceptable from with little food loss during the movement. This is the ultimate goal of any food supply chain and not increased food production alone.

Agricultural crops contain 65-95% water, and they continue their living processes post-harvest also. Their post-harvest life depends on the rate at which they use up their stored food reserves and their rate of water loss. When food and water reserves are exhausted, the produce dies and decays. Anything that increases the rate of this process may make the produce inedible before it can be used. The principal causes of loss are therefore discussed below, but in the marketing of fresh produce they all interact, and the effects of all are influenced by external conditions such as temperature and relative humidity.

Food losses after harvesting may include deterioration by biological or microbiological agents and mechanical damage due to unfavorable climate, cultural practices, poor storage conditions, and inadequate handling during transportation all of which can lead to accelerated product decay. Food losses also can either be due to the reduction in weight of food meant for consumption or it could be due to damage of physical spoilage (which is usually reported as a percentage of the food sample since it is difficult to measure it).

Improper harvest and post-harvest practices expose the food commodity to many deterioration agents which lead to food spoilage. There are various types of losses depending on the post harvest practices and deterioration agent. Losses are caused mainly by mechanical damage during transportation. Food losses lead to a loss or reduction in quantity, quality, nutritional and economic value of the food produce. These losses could either be primary, secondary or tertiary.

These are the losses that affect the food produce directly. They are during the food delivery chain. They include:

Microbiological Losses

These losses are as a result of the action of micro-organisms e.g. bacteria, mould and fungi. During the packing of vegetables, fruits and fishes into boxes, crates, baskets and trucks after harvesting, they are mostly subjected to cross-contamination by spoilage. These agents produce toxic substances (like mycotoxins) which causes food commodities to rot. These losses are more of loss in nutritional value than loss in weight. This occurs mostly during storage and marketing stages.

Chemical Losses

These losses are as a result of the reaction of the naturally present chemical constituents in the stored food to cause loss of color, flavor, nutritional value and texture.

Losses due to Biochemical Reaction

It refers to reactions of which intermediate and final products are undesirable. These can lead to significant loss of nutritional value such as rancidity and agro-chemical contamination and in most cases the whole vegetable, fruit and fish is lost. On the other hand, they are losses as a result of the reaction of chemical and biological constituents of the stored food. These losses give rise to discoloration and softening which leads to reduction of nutritional and economic value of the food product.

Mechanical and Physical Injuries

These losses are usually caused by bruises, cuts and excessive peeling of fruits. Mechanical damage is mainly due to in appropriate methods used during harvesting (careless handling), packing and inadequate transportation, which can lead to splitting, thus rapidly increasing water loss and the rate of normal physical breakdown. Skin breaks and other forms of mechanical damages also decay and promote the growth of pathogenic microorganisms.

Physiological Losses

An increase in the rate of loss because of normal physiological changes is caused by conditions that increase the rate of natural deterioration, such as high temperature, low atmospheric humidity and physical injury. Abnormal physiological deterioration occurs when fresh produce is subjected to extremes of temperatures, atmospheric modification or by contamination. This may cause unpalatable flavors, failure to ripen or other changes in the living processes of the produce, making it unfit to use. Physiological losses on the other hand, refer to the aging of products during storage due to natural reactions. They are as a result of the respiration of food products even after harvesting. This respiration causes a loss of weight and it produces heat which makes the food susceptible to micro-organism attack. Also physiological changes make the food product susceptible to mechanical damage.

Physical losses

These losses are mainly caused by the effect of temperature on foods. In closely confined storage, wrong environmental condition can result in microbiological losses.

Insect Pests And Parasites

These losses are as a result of the action of biological agents like rodents, insects, birds etc. the agents usually consume the food during storage and causes a reduction in weight and quality of the food. Fresh produce can become infected before or after harvest by diseases widespread in the air, soil and water. Some diseases are able to penetrate the unbroken skin of produce: others require an injury to cause infection. Damage so produced is probably the major cause of loss of fresh produce.

Secondary Losses

These are losses that do not affect the produce directly, but presents favorable conditions for the actions of primary losses. They are incurred during the delay in food processing and delivery chain. They are usually as a result of inadequate harvesting, packaging, transportation, storage and drying or processing facilities and poor quality control practices.

Tertiary Losses

These losses are usually caused by the consumer due to unhygienic and careless handling of the foodstuff which can lead to wastage or loss. Various surveys have been carried out to assess the losses of food crops. A qualitative assessment must be made in order to know the post-harvest practices to prevent huge losses. The type of measures required to reduce the losses and the manner in which the measures should be adopted and applied also needs to be judged properly and with adequacy. A general assessment for food commodities cannot be made because the losses for different commodities differ significantly. The losses depend on the harvest and post-harvest practices which depend on the type of produce, final use, climate, harvesting practices and the social and cultural practices of the consumers.

Post-harvest Damages

Physical damages to fresh produce can come from variety of sources, the most common being:

The high moisture content and soft texture of vegetables, fruits and fishes make them susceptible to mechanical injury, which can occur at any stage from production to retail marketing.

Injuries from Temperature Fluctuations

All fresh produce is subject to damage when exposed to extremes of temperature during chilling and freezing. Commodities vary considerably in their temperature tolerance. Their level of tolerance to low temperatures is of great importance where cool storage is concerned.

To prevent and reduce the losses of harvested produce it should be handled and stored hygienically under conditions of optimum temperature and humidity. Processing techniques under controlled conditions also proves to be useful in this regard. This improves the shelf-life of the stored agricultural produce until it reaches the consumers.

Chapter 2
Food Contamination and Spoilage

Food is liable to be contaminated at any stage from producer to consumer. Mostly all outbreaks of food poisonig occur during the summer months. In many European countries, food poisoning is not properly reported at times. is entirely There veterinarians are generally vested with the duty of food inspection.

The significance of milk as human nutrition is now well established as it is considered as the best ideal and complete food for all age groups. In spite of all these factors, milk can also serve as a potential vehicle for transmission of certain diseases under circumstances. Moreover by virtue of possessing almost all the essential nutritional factors, milk can also serve as an excellent culture and protective medium for certain micro organisms which may include potential pathogens capable of causing serious health problems. Milk can also allow some organisms to grow and produce certain toxins there by making itself a vulnerable commodity from public health point of view. As per WHO and FAO the milk borne diseases can be classified in two groups:

1) Infection of animals that can be transmitted to man through milk.

2) Infection primarily of man that can be transmitted through milk.

The health of dairy animals is a very important consideration because a number of diseases of cattle including Brucellosis, Q fever, Salmonellosis, *Staphylococcal* and *Streptococcal* infections and Foot and Mouth Disease (FMD) virus may be transmitted to man through the consumption of milk. The organisms of most of these diseases may be transmitted to milk either directly from the udder or indirectly through the infected body discharges, which may drop, splash or be blown into milk.

The diseased persons may transmit disease like typhoid fever, scarlet fever, diphtheria, septic sore throat, and infantile diarrhea by contaminated hands or by coughing, sneezing and talking.

Dairy and farm environment may also introduce a variety of pathogens into milk and milk products at different stages of production and processing. Some common air-borne pathogens include Group A *streptococci, Corynebacterium diphtheriae, Mycobacterium tuberculosis, Coxiella burnetti* and some viruses of respiratory origin. Water, fodder and unhygienic conditions at farm and plant level may also contribute pathogens to milk.

Prevention of milk-borne diseases is one of the most important concerns of public health significance. Success in controlling a disease can be maintained only by the constant vigilance over the health of the cow udder until it reaches the mouth of the consumers.

Commonly Occurring Food-borne Infections of Public Health Significance

Salmonellosis: There are a group of bacteria of more than 2300 types or species commonly known as Paratyphoid Bacteria. They inhabit in the intestine of clinically healthy man and animal and also pathogenic for men or animals or both. Salmonella can exist in feces or on pastures for considerable period. They are not destroyed in carcass or offal maintained at chilling or freezing temperature. Salmonella can also grow well on meat at ordinary temperature. It is probable cause of 75% of outbreaks of food poisoning *Salmonella typhimurium* is found in rat, mice, cattle, sheep, goat, pig fowl and duck. *S. enteritidis* also cause meat food poisoning and it is commonly found in rat, cattle, pig, goat and duck.

Typical food poisoning commences within 7-72 hours after ingestion of the organisms. The foods involved is generally egg, meat and milk and is often derived from infected animals, but may also contaminated during storage or preparation. Salmonella food poisoning is characterized by nausea, vomiting, chills, abdominal pain and diarrhea. Salmonella must be controlled during processing and manufacturing of animal feeds, by frequent disposal of animal excreta in livestock farms and by transportation of livestock in convenient manner to reduce stress.

Campylobacteriosis: Campylobacteriosis or *Campylobacter enteritis* is caused by consuming food or water contaminated with the bacteria *Campylobacter jejuni. C. jejuni* commonly is found in the intestinal tracts of healthy animals and in untreated surface water. Raw and inadequately cooked foods of animal origin and non-chlorinated water are the most common sources of human infection. Diarrhea, nausea, abdominal cramps, muscle pain, headache and fever are common symptoms. Onset usually occurs in 2-5 days after eating contaminated food. Rarely deaths have also been reported.

Preventive measures for *Campylobacter* infections include pasteurizing milk, avoiding post-pasteurization contamination, properly cooking meat and allied products.

Listeriosis: Listeriosis primarily affects newborn infants, pregnant women, the elderly and those with compromised immune systems. In a healthy non-pregnant person, listeriosis may occur as a mild illness with fever, headaches, nausea and vomiting. Among pregnant women, intrauterine or cervical infections may result in spontaneous abortion or still birth. Infants born alive may develop meningitis. The mortality rate in diagnosed cases is 20-35%. The incubation period is a few days to several weeks. Infection is usually derived from meat, poultry and fish products.

Preventive measures for listeriosis include maintaining good sanitation, turning over refrigerated, pasteurization and avoiding post-pasteurization contamination of food products.

Staphylococcal intoxication: *Staphylococcus* spp. of bacteria are found on the skin and in the nose and throat of most people; people with colds and sinus infections are often carriers. Infected wounds, pimples, boils and acne are generally rich sources. *Staphylococcus* also is widespread in untreated water, raw milk and sewage. When *Staphylococcus* bacteria get into warm food and multiply, they produce a toxin or poison that causes illness. The toxin is not detectable by taste or smell. The bacteria itself can be killed by temperatures of 120°F. Symptoms include abdominal cramps, vomiting, severe diarrhea and exhaustion. These usually appear within 1-8 hours after eating staph-infected food and last one or two days. The illness seldom is fatal. Foods commonly involved in staphylococcal intoxication include protein foods such as ham, processed meats, tuna, chicken, sandwich fillings, cream fillings, potato and meat salads, custards, milk products and creamed potatoes. Foods that are handled frequently during preparation are prime targets for staphylococci contamination.

Clostridium perfringens **food borne illness:** *Clostridium perfringens* belong to the same genus as the botulinum organism. However, the disease produced by *C. perfringens* is not as severe as botulism and few deaths have occurred. Spores are found in soil, nonpotable water, unprocessed foods and the intestinal tract of animals and humans. Meat and poultry are frequently contaminated with these spores from one or more sources during processing. Spores of some strains are heat resistant and can survive boiling for four or more hours. Furthermore, cooking drives off oxygen, kills competitive organisms and heat-shocks the spores, all of which promote germination. Sufficient numbers of vegetative cells may be produced by the bacteria under conducive conditions to cause illness. Foods commonly involved in clostridium illness include cooked meat and processed poultry products. Symptoms appear within 8-24 hours after contaminated food is consumed. They include acute abdominal pain and diarrhea; nausea, vomiting and fever are less common. Recovery usually is within one to two days, but symptoms may persist for one or two weeks.

E. coli hemorrhagic colitis: *Escherichia coli* belong to a family of microorganisms called coliforms. Many strains of *E. coli* are saprophytes in animal gut. However, *E. coli* O157:H7 strain causes a distinctive and sometimes deadly disease. Symptoms begin with nonbloody diarrhea one to five days after eating contaminated food, and progress to bloody diarrhea, severe abdominal pain and moderate dehydration. In adults, the complications sometimes lead to thrombocytopenic purpura (TPP) characterized by cerebral nervous system deterioration, seizures and strokes. Minced beef is mostly associated with *E. coli*

O157:H7 outbreaks, but other foods which include raw milk, unpasteurized foods not exposed to heat trteatment and untreated water are also implicated. Infected food handlers with the disease likely help spread the bacteria.

Preventive strategies for *E. coli* infections include thorough washing and other measures to reduce the presence of the microorganism on raw food, thorough cooking of raw animal products, and avoiding recontamination of cooked meat with raw meat. To be safe, cook ground meats to 160°F.

Viruses: Nearly one-third of food poisining incidences worldwide majorly in developed countries is believed to be viral in origin. In USA, viral food intoxications contribute to overall 50% cases annually including norovirus as prominent source, which individually share 57% of the total outbreaks in this regard. The severity of virus borne food intoxications have intermediate incubation period causing illnesses which may be self limited or have the potential to spread among healthy individuals. Common viral etiology includes Enterovirus.

Hepatitis A causes viral illnesses through the carrier of food lasting for 2-6 weeks incubation period and generally characterized by its systemic nature of spread throughout the system of individual. It also leads to liver damage and dysfunction through causation of jaundice. Liver damage caused can be acute or chronic in nature. The infection can originate from consumption of fresh products or those which have been exposed to fecal contamination (Dubois *et al.*, 2007; Schmidt, 2007).

Apart from this other viral source include Hepatitis E, Norovirus and Rotavirus as significant in this course.

Parasites: These also contribute significantly towards causing food borne illnesses. Most of the food borne intoxicoses by parasites is zoonotic in nature. These include Platyhelminthes: *Diphyllobothrium* sp., *Nanophyetus* sp., *Taenia saginata*, *T. solium*, the scolex of *T. solium*, *Fasciola hepatica* along with tapeworm and flatworm.

Nematode: *Anisakis* sp., *Ascaris lumbricoides*, *Eustrongylides* sp., *Trichinella spiralis*, *Trichuris trichiura*.

Protozoa: *Acanthamoeba* and other free-living amoebae, *Cryptosporidium parvum*, *Cyclospora cayetanensis*, *Entamoeba histolytica*, *Giardia lamblia*, *G. lamblia*, *Sarcocystis hominis*, *S. suihominis*, *Toxoplasma gondii*.

Fish and aquatic sources: These include shellfish toxin, including paralytic shellfish poisoning, diarrhetic shellfish poisoning, neurotoxic shellfish poisoning, amnesic shellfish poisoning and ciguatera fish poisoning, Scombrotoxin, Tetrodotoxin (fugu fish poisoning).

Natural toxins: These kinds of toxins are contained in several food products which are natural in origin and are not derived from bacterial source. Animal flesh sometimes is rendered toxic from the products derived from plant sources in their feed. Plant toxins generally include alkaloids, Ciguatera poisoning, Grayanotoxin (honey intoxication), mushroom toxins, Phytohaemagglutinin (red kidney bean poisoning; which can be made inactivated by boiling) and Pyrrolizidine alkaloids which can infect food sources causing food borne toxicity to human beings. Some substances originating from plants are therapeutic in regulated dose but prove to be toxic in high dosage. Foxglove, derived from plant source, contains cardiac glycosides; Poisonous hemlock (conium) has medicinal uses.

Other pathogenic agents: These include non-conventional intracellular virus like agents called Prions, resulting in causation of Creutzfeldt-Jakob disease (CJD) in human.

Ptomaine poisoning: Another predominant food toxicity called *"Ptomaine poisoning"* which involves *ptomaines* (from Greek *ptôma*, "fall, fallen body, corpse") which are alkaloids obtained in decomposed and decaying vegetable and animal matters. However, the disease causing bacteria causing food borne illnesses do not produce ptomaine as toxin and so not used or scientific purposes nowadays in the field of zoonotic food borne toxicity.

Emerging foodborne pathogens: Many emergent foodborne illnesses are new to the field and are not completely understandable till date. Approximately 60% outbreaks are caused by new and emergent unknown sources derived from *viz., Aeromonas hydrophila, A. caviae and A. sobria.*

Control of Food Poisoning

The food should be properly refrigerated for preventing food poisoning. Strategic rules for hygienic preservation of food for benefit to public health should be employed along with proper and strict survey of animal products from source to transformation industry and delivery points. The regulation of food quality should include traceablity of the source of food poisoning in final processed product for which it is needed o know the origin an dquality of each ingredient i.e., keeping track record or history oif the animal from which it is derived and assessing the efficiency of processing technique. In this way the origin of any probable food poisoning can be tracked and regulatory measures for its control and prevention in processed products can be implemented. Enforcement of strict hygienic layouts like HACCP and preservation of perishable products through cold chain technique should be adopted. Veterinarians also have apivtal role to play in this entire process by promulgating enforcement laws.

Under domestic conditions, prevention of food intoxications and food safety practices generally include sufficient cooking at optimum temperature and followed by its prompt and effective refrigeration. However, some bacterial exotoxins released in food are resistant to heat treatment.

Infection of Animals Transmitted to Man through Milk

The diseases in principle concerned in milk hygiene are TB, Brucellosis, Streptococcal infection, Staphylococcal enterotoxin poisoning, Salmonellosis and Q-fever. Diseases of lesser importance include cow pox and vaccinia, pseudo cowpox (Milker's Nodule) which are usually transmitted to milkers through contact during the act of milking rather than through ingestion of milk. FMD, Anthrax and Leptospirosis have also been transmitted on rare occasions. The tick borne Encephalitis virus may also be transmitted through milk but further is required, although it is theoretically possible that the polio virus infection and other enteroviruses to be transmitted through milk. However, milk borne hepatitis has been recorded in several times (occasions). The organism that causes all the diseases as mentioned above will be inactivated except the spores of *Bacillus anthracis* by adequate pasteurization or heating.

Tuberculosis: Human beings are universally susceptible to infection with the bovine tubercle bacillus. Bovine TB is particularly common in children. It is the most common pathogen present in raw milk. The milk from single infected cow might contaminate the entire milk supply through mixing. The bovine tubercle bacillus is chiefly responsible for the non-pulmonary type infection in man. Children become infected most commonly in the alimentary tract, cervical lymph glands, bones and skin. Organisms enter milk directly from udder or from dung. Pasteurization kills the pathogen. Tuberculosis has been recognized as one of the most important infectious diseases in the world because of its high global impact and its chronic debilitating characters particularly in the poor sections of the society. Milk borne tuberculosis is directly or indirectly related to the consumption of raw milk from infected dairy herd. Tuberculosis of milk producing animals has two significant features:

i) Cattle and goats infected with any of the three types of tubercle bacilli *i.e. Mycobacterium tuberculosis, M. bovis* and *M. avium viz.* human, bovine and avian types which excrete the organism in the milk even though their udder may be distinctly normal. For this reason, any animal that is known or suspected to be infected with tubercle bacilli or that reacts to the tuberculin test that should be considered as a potential actual excreter of the organism regardless of the presence or absence of signs of udder abnormalities.

ii) The human or bovine strains of the tubercle bacilli may be transmitted between animals and men who are in direct contact with them not only through the ingestion of infected materials but also through the respiratory tract.

Prevention and control: All cattle should be subjected to Tuberculin test to find out positive or infected cases. Tuberculosis cases should be eliminated. Proper pasteurization or heat treatment of milk and milk products should be practiced. The attendants who are positive for tuberculin test should be prohibited from handling the cattle or milk. Overcrowding in the farms should be avoided. Animal house or living conditions of persons or attendants should be improved with proper ventillation. Proper disinfection of udder clothings, utensils for milk collection should be disinfected properly. Handlers should be subjected to 'Mantoux test' to find out whether that particular person is affected with tuberculosis.

Brucellosis: Infection is primarily acquired from infected animals. The species involved are *Brucella melitensis, Brucella abortus,* and *Brucella suis.* Man is infected directly by contact with the diseased animal or their tissue or discharges or through consumption of raw milk and milk derivatives or diseased animal. The organisms get established in the udder and multiply there and are excreted in the milk in large numbers. *Brucella abortus* can also get into the milk from unsanitary barns as the organism survives for long periods under a variety of conditions. The safest method of disease prevention is to discard the affected animal's milk. Pasteurization is effective in destroying the microorganism.

Prevention and Control: Segregation of infected herd should be done to avoid cross infection of milk of the healthy herds. The healthy animal should be isolated and slaughtered. The herd should be properly vaccinated with calf-hood vaccination by using Brucella cotton strain-19 vaccine. Vaccine should be administered to calves below 6 months of age. Adequate heat treatment and pasteurization of milk and milk products should be done. The persons engaged in dairy farms should be tested against Brucellosis.

Anthrax: Anthrax disease has been reported in people following consumption of infected milk. *Bacillus anthracis* may be excreted in milk shortly before death of the animal. Samples of milk drawn from cows that died of Anthrax may show numerous anthrax bacilli. However, the organisms may gain entry into milk from environment which is often contaminated with the bacilli or its spores.

Prevention and Control: Milk from disease animals should not be used. Proper sanitation of the dairy herds, proper vaccination to the animal of dairy herds by Anthrax spore vaccine (1 ml dose) should be implemented. The discharges of the infected materials should be promptly and properly disposed off.

Q fever: It is caused by *Coxiella burnetii* and is a rickettsial disease. The organism is known to be excreted in the milk of cattle, sheep and goats. Raw milk is commonly implicated as a vehicle for the transmission of the disease. A number of reports indicating the high incidence of disease in raw milk as well as pasteurized milk have been recorded from different countries. *C. burnetii* is more resistant to heat and it can resist at certain pasteurization temperature i.e. Vat pasteurization temperature especially when air space heaters are not used, it may escape destruction at low temperature accepted for pasteurization i.e. 61.5°C for 30 min, thus leading to potential hazard. Treatment at 63°C for 30 min or 75°C for 15 sec have been recommended for the inactivation of organisms in milk as *C. burnetii* is comparatively resistant to heat than *M. tuberculosis*. Higher temperature is recommended for milk and milk products where butter fat content is greater than that of milk.

Prevention and Control: (i) Adequate heating of milk and cream, (ii) Calving and lambing sheds should be away from the milking shed and dairy farms, (iii) Heat treated milk should be kept away from dust and discharges, (iv) Control on the import of domestic animals and (v) Recontamination of heat treated milk with dust and discharges should be prevented and avoided.

Cow Pox: Cow Pox is transmissible from infected cows to the human beings, especially to the young one. In the case of adults, the attack will be usually mild but in children, it results in fever and malaise. The contamination of milk is through the lesions on the udder. Pasteurization is effective in destroying the virus.

Prevention and Control: The veterinary advice on herd treatment should be short and the milk should be pasteurized properly. The milk of cows showing active lesions in the udder in the form of pustules or serous scabs on the udders should not be used. The principle sources of infection for milkers are handling the teats of such infected animals.

Foot and Mouth Disease: This virus produces gastro-intestinal disturbances on consumption of milk and dairy products like cream, butter and cheese prepared from milk of infected animals. Children are more susceptible. The virus is present in the fluid of the vesicles and from these vesicles; they gain entry into the saliva, feces, urine and milk. Lesions also occur in the udder, milk ducts and secreting portions of the udder is also involved. Pasteurization is a simple method to destroy the FMD virus.

Staphylococcal enterotoxin intoxication: Multiplication of certain strains of Staphylococcus in milk leads to enterotoxin formation in milk. This toxin is particularly produced by enterotoxic strains of *Staphylococcus* spp. These strains show the usual characteristics of pathogenic staphylococci and are coagulase positive in nature. Enterotoxin contamination of raw milk is usually derived from sick animals, usulay who suffer from mastitis. It can be from food handlers or milk handlers who carry the bacterial population on their skin in hand and forearms in conditions such as pyodermitis and furunculosis. Staphylococcus enterotoxin which generally remains preformed in milk is thermostable in nature and can resist boiling.

Prevention and control: (i) Rapid refrigeration of food or milk or milk products, (ii) Control of human carriers at critical points in milk processing plants and (iii) maintenance of hygienic standards in milk processing plants during transport and up till distribution to consumers.

Infection Primarily of Man Transmitted through Milk

Typhoid or enteric fever: It is caused by *Salmonella typhi*. Milk forms the chief source of infection, which occurs through infected persons (active cases) and carriers are potent sources. Use of contaminated water for washing the utensils is another source. Infection through flies is possible. Water obtained from trusted sources largely eliminates the chances of typhoid organisms entering into milk through water. Elimination of human carriers is very important. Pasteurization of milk invariably kills the bacteria.

Paratyphoid: The etiological agent is *Salmonella paratyphi* and spread of this disease through milk and dairy products show similar features as typhoid and the sources of infection are the same as in typhoid. The paratyphoid bacilli are destroyed when the milk is heated to a temperature of 59°C for 10 min.

Shigellosis or Bacillary dysentery: Shigellosis is one of the common food borne infections caused by *Shigella dysenteriae*. Milk borne outbreaks have been frequently recorded and unpasteurized milk appears to be commonly implicated in such outbreaks. *Shigella* multiplies readily in milk at a temperature of 15°C or above. Milk may become infected by contamination with infected materials like utensils, water, flies, etc. Milk handlers may be carriers of infectious agents and also cause contamination.

Prevention and control: Rigid discipline of proper sanitation should be enforced among dairy workers particularly in pasteurization plants and retail shops dispensing milk in bulk. All the sanitary attendants looking after patients should be prohibited from contact with milk or utensils. Fly population should be controlled adequately during milk production and distribution.

Cholera: Outbreaks of this disease due to infected milk have been reported in various parts of the world. The infection of milk may be through carriers or through infected water used for adulteration. Human carriers have been proved to be very potent source of infection of the milk. The causative microorganism *Vibrio cholerae* apparently is easily destroyed by acid development in milk and cream.

Prevention and control: Prevention consists of periodic medical inspection of all persons connected with farm operations and pasteurization of milk, as the organism is destroyed at 55°C.

Streptococcal infections: Streptococcal infections like septic sore throat, scarlet fever and food poisoning have been traced to the consumption of contaminated milk and dairy products. Few strains of group D streptococci and enterococci have been found to produce toxic metabolites in milk.

Streptococcus pyogenes: This bacterium causes scarlet fever, septic sore throat, tonsillitis and septicemia in humans.

Streptococcus *agalactiae:* It is the causative agent for mastitis in animals.

Group D *streptococci:* It is one of the pathogen behind incidence of food poisoning in the human beings.

Scarlet fever is primarily a disease of man, but cows may be secondarily infected from human sources and pass on the disease through milk to man. Large number of streptococci shed from udder of single cow infected with this organism may contaminate a large bulk of milk. Pasteurization of milk is the only safeguard against these bacteria.

Prevention and control: (i) Adequate heat treatment of milk and its products, (ii) Holding of milk at low temperature during storage, (iii) Rejection of milk from animals suffering from mastitis, (iv) Regular check on the health of the dairy workers and (v) Fecal contamination of milk and its products should be avoided.

Diphtheria: Milk borne epidemics of diphtheria have been reported from lesions on the teats from which the organisms were isolated. However, the primary sources of the organism have been traced to either the milk man or some farm worker. The organism produces an extra cellular toxin in the respiratory tract of man, which sometimes causes death. The animals got infected secondarily, and the human carrier is an important factor.

Prevention and control: Prevention consists of rigid medical inspection of all dairy workers and pasteurization of milk makes it safer for human consumption.

Listeriosis: Listeriosis is a food-borne illness caused by *Listeria monocytogenes*, is capable of growing over a wide range of temperatures from 1-45°C. The pathogenicity is due to the production of extracellular haemolysin (alpha, beta types) by this organism. Infected animals are the main sources of these organisms. Handlers carrying the disease, unhygienic practices during production and processing, faecal contamination of milk and water, contaminated refrigerators, dish and clothes may also introduce the organisms into milk. This organism appears to survive heat treatments like pasteurization. The possible reason for the survival of this organism is the ingestion of these organisms by leukocytes in milk, which may give some protection against heat treatment. Ingestion of *L. monocytogenes* through milk and milk products can cause listeriosis in man leading to the death of the consumer. The symptoms of listeriosis are acute meningitis with or without septicemia.

Prevention and control: (i) Strict hygienic practices should be followed during production and processing, (ii) Prevention of source of infection and (iii) Proper heat treatment of milk.

Bacillus cereus poisoning: *Bacillus cereus* is one of the causative organisms for mastitis. The herd raw milk sometimes gets *B. cereus* from mastitic animals. The raw milk also gets spores from animal's teats and skin, milking machine and other source such as cans. Soil may introduce *B. cereus* directly or indirectly into milk and milk products. Three types of toxins are elaborated by toxigenic strains of *B.cereus* in milk and milk products which are Haemolysin, Lecithinase and Enterotoxin. Enterotoxin is responsible for food poisoning outbreaks, while lecithinase and haemolysin play important role in the pathogenesis of the organism.

Prevention and control: (i) Prompt cooling of milk/ product during storage and (ii) General hygienic conditions should be maintained during production of milk and marketing of dairy products.

Aflatoxicosis: Aflatoxicosis is common type of fungal intoxication caused by the common moulds *Aspergillus flavus* and *A. parasiticus* by virtue of their ability to produce aflatoxin. *A. flavus* can produce aflatoxin B1, B2 and G1, G2 types in milk and milk products. Upon ingestion, the *Aspergillus* toxins are metabolized by the milch animals and are secreted into the milk in the form of M1 and M2, which are also toxic to the milk human beings who consume the contaminated milk. These toxins are extremely heat stable, potent and exhibit very strong toxicity apart from being highly carcinogenic. Aerial contamination is one of the most important sources of mould spores. Soil and contaminated food may also introduce spores in milk and milk products. Poor storage conditions, especially damp weather favur toxin production.

Prevention and control: (i) Preventing fungal contamination of milk and milk products as well as feed by taking appropriate precautions. (ii) Preventing fungal growth of the animal feeds by storing the products under proper conditions and by use of fungistatic agents and (iii) By detoxification of aflatoxins by physical and chemical agents.

Rabies: Rabies virus has been found to be excreted in the milk of infected animals, although the spread of this deadly disease through milk is rare. It has been found that milk from rabid animals is avirulent and is not capable of producing the disease. It is also reported that there is no danger in the ingestion of infected milk unless there are any abrasions on the lips or along the gastrointestinal tract. However, as a precautionary measure, milk from suspected cows should not be used for human consumption. Pasteurization is effective in destroying the virus.

Actinomycosis: The possibility of infection of human beings with *Actinomycosis bovis* is certainly remote, though not impossible. The microorganisms if present in milk, can affect the human body only by gaining entrance through raw wound, such as a new tooth cavity or intestinal ulcers etc. Milk should be discarded or at least pasteurized.

Digestive disturbances: Food poisoning cases have been reported following consumption of milk from cows suffering from intestinal disorders and the causative microorganisms might have reached the milk through feces. Pasteurization is effective in destroying most of the pathogenic microorganisms.

Chapter 3
Microbial Diversity as Pathogens in Food

Studies have shown that one of the most common types of food intoxication is caused by certain staphylococcus strains, mainly *Staphylococcus aureus* (Jablonski and Bohach, 2001; Evenson *et al.*, 1988). The natural habitat for *Staphylococcus* spp. being the skin and mucus membranes of animals and men, it does not multiply in fish and its presence in fish indicates post harvest contamination (Huss, 1994). *S. aureus* is an indicator of hygiene and sanitary conditions the presence of this organism indicates the unhygienic condition during processing, storage etc and the contamination of fish could be the result of combination of improper handling, improper storage and cross contamination (Simon and Sanjeev, 2007).

The presence of coliforms, especially *E. coli*, is an indication of fecal pollution and is considered as an indication of the presence of other pathogenic bacteria. *E. coli* counts more than 100 per gram, can be considered as heavily contaminated with fecal matter. The poor quality of ice used supplements to this contamination (Sudhi, 2002). The natural habitat of *E. coli* is the gastrointestinal tract of warm blooded animals. ICMSF (1986) recommends testing for the presence of this organism as an indicator of post harvest contamination particularly from fecal origin and its limit is recommended as <100 *E. coli* g^{-1} fish. The abundance of *E. coli* has been shown to be more related to the sanitary risk than that of coliforms. Sudhi (2002) investigated it in fish of Kochi market (Kerala) and Nair and Nair (1988) in *Labeo rohita* and *L. calbasu*.

Salmonella and other bacteria may contaminate seafood during processing, and may cross-contaminate products during various stages of preparation (Amagliani *et al*. 2011). High count of *Salmonella* seems to be due to unhygienic handling, improper method of icing, exposure to outer environment such as soil, filthy water and surfaces etc. *Salmonella* and other bacteria may contaminate seafood during processing, and may cross-contaminate products during various stages of preparation (Amagliani *et al*., 2011). The result of this finding can be compared with the results of Sinha *et al*. (1991) in marketed rohu (*L. rohita*); Jeyasekaran and Ayyappan (2003) in rohu (*L. rohita*).

The vibrios remain present as predominant bacteria of estuarine water (Panda and Nayak, 2001). Subburaj (1984) reported that the market premises and market floor and that water could be major sources of contamination of fish in the fish markets in Mangalore, India.

The normal range of Total plate count (TPC) in the intestine of fresh water fish had been reported as to be 10^3-10^9 by Shewan (1962). The gill is an ideal site for microbial growth (Russel and Fuller, 1979). Shewan and Murray (1979) reported that in case of chilled fish only few bacteria invade the muscle at the late stage of storage. Suhendan *et al*. (2007) studied psychrophilic bacteria in fish and concluded that the estimation of the psychrophilic microorganisms could be a better tool for shelf-life estimation of chilled fish than that of mesophilic bacteria.

The natural habitat for *Staphylococcus* spp. being the skin and mucus membranes of animals and men, it does not multiply in fish and its presence in fish indicates post harvest contamination (Huss, 1994). Presence of *Staphylococcus* spp. in high numbers indicates unhygienic sanitary condition of the market. The incidence of *Staphylococcus* sp. in the fishery products within limits is not a serious problem. However, careless handling during processing results in the multiplication of the organism, which may lead to food poisoning (Lilabati and Vishwanath, 1999). High counts of *Staphylococcus* spp. have also been recorded in *L. rohita* at two different temperatures (Jeyasekaran and Ayyappan, 2003) and Kumari *et al*. (2001) in *L. rohita* available at Patna market.

Chapter 4
Food Processing

There are several advantages of proper food processing under controlled and regulated conditions. It implies the decrease or removal of the content of anti-nutritional factors from the food, increase in shelf-life for prolonged preservation, ease in marketing and increase in consumer demands and increment in the quality and consistency of the finished processed food. It also increases the availability of many food items during off-seasons, increases the convenience in transportation of food items over long distances by decreasing the chances of rotting of mainly perishable food items and increasing the safety for consumption by deleting pathogenic microorganisms which cause spoilage. Food processing at certain places can also be used to reduce the conditions of food shortage and by supplementation of nutritious and safe food for the masses.

Fermentation is a microbial technique and the reaction to be controlled in favorable and desirable conditions for food safety and quality after fermentation, especially in the production of alcoholic premium quality beverages like beer, wine and cider. The same technology is employed in the bred manufacturing industries for leavening activity brought about by the production of carbon dioxide by the microbial or yeast activity. The preservation effect during fermentation is attributed to the production of lactic acid in sour foods such as yoghurt, dry sausages, pickles, sauerkraut and vinegar (extremely diluted acetic acid).

The fermentation technology under controlled conditions is an age old practice both in households and industries for food processing and preservation, be it alcoholic beverage products of edible products derived from vegetable, fish and meat sources. Louis Pasteur, the renowned French chemist is the world famous and first known zymologist in history, who in 1856 established the pivotal role of yeasts in fermentation. Pasteur originally defined fermentation as "respiration without air" after regular performances of lengthy experimental protocols. After observation of the breakdown of sugars to alcohols by the action of yeast, the pioneer concluded that the entire reaction is driven by the chemical catalytic action of certain forces called ferments inside the yeast cells. It was further observed that the yeast extracts can bring bout fermentation of sugars even also in the absence of viable yeast cells. In 1897, Eduard Buchner of Humboldt University of Berlin, Germany discovered that sugars are fermented in the absence of viable cells also in the fermentation mixture. The yeast cells secrete a chemical component called zymase. For his memorable contributions in research and discovery of cell-free fermentation, in 1907 Buchner was awarded with the prestigious Nobel Prize in Chemistry. In 1906, NAD$^+$ was discovered out of studies carried out from ethanol fermentation.

Food is generally exposed to ionizing radiations for destroying microorganisms, bacteria, viruses, or insects which the food might contain. Further applications include delay of ripening, increase of juice yield, sprout inhibition and improvement of rehydration. In certain cases, food irradiation leads to substantial chemical changes.

Irradiation is effective not only on food items but also on non-food items, such as medical hardware, plastics, tubes for gas pipelines, hoses for floor heating, shrink-foils for food packaging, automobile parts, wires and cables (isolation), tires, and even gemstones.

Food irradiation is very promising as a new technique in enhancing food safety and quality standards by causing the destruction of pathogenic microorganisms such as *E. coli* O157:H7, *Campylobacter*, and *Salmonella* from foods·

Irradiation process also helps in reduction of spoilage bacteria, insects and parasites. It is gained its importance as a high quality food hygiene practice by causing the reduction and elimination of harmful and enteric microorganisms and bacterial population. The Food and Drug Administration has approved irradiation as an effective food quality technique for preservation and increasing storage life of meat, fresh fruits, vegetables and spices. Irradiation process is also used in certain fruits and vegetables for delaying and inhibiting sprouting and ripening processes.

The effects of irradiation on the food and on animals and people eating irradiated food have been studied extensively for more than 40 years. These studies show clearly that irradiation process is approved for application on foods: Food irradiation is a very efficient mechanism in prevention of many food borne diseases and intoxications.

Food preservation by irradiation technique provides consumers with wholesome and nutritious food items having improved hygiene and easy availability and quantity with increased storage life, convenience to transport. So, the process adds cost to fresh and preserved food items.

Benefits of Processing

As food processing decreases the population or load of pathogenic microorganisms in food and neutralizes the harmful mycotoxins, if present therein. So, it reduces the chances of food-borne diseases caused by microorganisms like Salmonella etc. which can harbor in raw meat and incidences of mycotoxicoses (majorly, aflatoxicosis, ochratoxicosis and zearalenone) due to prolonged improper storage of food thereby causing human illnesses. Food processing has also gained its importance in the wide variety of diet among people throughout the globe and availability of exotic food items at various places. Processing of food items enhance the taste, flavor and aroma of the food thereby increasing the overall chances of its acceptability among the masses.

Food processing whenever performed in large mass is comparatively cheaper than processing and modification of individual ingredients. So, the food processing sector implies a huge margin of profit for processed food manufacturers and retailers in the supply chain.

Processing involves various methods among which cooking is a very popular and widely used method which involves the modification by blending etc. of naturally available unprocessed food ingredients. In our nowadays fast paced lifestyle where every family member is on a go for financial security, processed food products have gained its important position in daily livelihood by offering ready prepared wholesome and nutritious meals within short period.

The modern methods of processing decreases the risk of health hazards to consumers from diabetics, allergies etc. Food processing also involves fortification for the production of neutraceuticals and energy supplements with addition of probiotics, prebiotics, certain important vitamins and mineral elements within standard permissible limits which are rather present in natural food in very scarce quantity.

Drawbacks and Limitations

There exist certain limitations of food processing also. For example, during processing by heating the concentration of vitamin C is reduced, as it is heat-sensitive. Generally, food processing techniques reduce the nutritional quantity in very negligible amount of nearly 5-20%. Food processing involves the use of food additives, which sometimes prove to be detrimental to public health. For this reason, the European Food Safety Authority (EFSA) has specified the level of individual feed additive during processing technique and which is approved for safe consumption of human beings. The additives after approval gain an 'E' number (E stand for Europe) which signifies the quantity of the additive to be incorporated in the finished processed food item.

Food processing involves many mechanisms like mixing, grinding, chopping and emulsifying during the whole process of production, which indirectly increase the chances of contamination and admixtures with undesirable foreign elements. Sometimes, packaging containers also pose a threat for public contamination when exposed to thorough procedures of continuous processing by leaching of the chemical components from the containers into the food item to be processed.

In food manufacturing practices, using metal detectors decrease the risk of contamination with metal fragments during the processing technique. In large food processing equipments are fitted with many metal detectors at several positions to negate the chances and risks of metal contamination of processed food products. In 1947, the first industrial purpose metal detector was introduced by Goring Kerr.

Fermentation technology is primarily employed for the preservation of different food by production of acids and alcohols, biological fortification and enrichment of food items with potential biogenic products like essential amino acids, easily digestible proteins, essential fatty acids and useful vitamins, neutralization of anti-nutritional factors, to diversify and enrich the diet with various aromas, flavours and textures in food substrates and decrease in requirements of further processing techniques like cooking etc.

Specifically, in the fish processing technological research aspect, Bagoong, Faseekh, Fish sauce, Garum, Hákarl, Jeotgal, Rakfisk, Shrimp paste, Surströmming,Shidal and Ngari are the popular fermented fish products worth mentioning.

Risk Elements Involved for Public Health

There are certain risks and health hazards associated with excess and regular consumption of fermented food products. In Alaska, since 1985, there has been increase in incidences of botulism exceeding the case reported in the Americas. This is mainly caused for the practice of allowing whole fish, fish heads and meat of animals like sea lions, walrus, whale flippers, birds, seal tallow, beaver tails etc. to ferment for prolonged periods before consumption by the resident Eskimos there. During this extended fermentation, if plastic wrappers or containers are used, then *Clostridium botulinum* gets a conducive condition to thrive in the micro-aerophilic condition inside the plastic containers.

Alaska has witnessed a steady increase of cases of botulism since 1985. It has more cases of botulism than any other state in the United States of America. This is caused by the traditional Eskimo practice of allowing animal products such as whole fish, fish heads, walrus, sea lion, and whale flippers, beaver tails, seal oil, birds, etc., to ferment for an extended period of time before being consumed. The risk is exacerbated when a plastic container is used for this purpose instead of the old-fashioned, traditional method, a grass-lined hole, as the botulinum bacteria thrive in the anaerobic conditions created by the air-tight enclosure in plastic containers.

Processing techniques can be lengthy and time consuming sometimes depending on the type of food being processed and it needs the control and regulation of certain parameters for processing which includes hygiene which is assessed by the microbial load in the processed food product, efficiency in energy utilization, minimum waste generation, effective labour saving and minimization of cleaning requirements. Hygiene protocols for the finished processed food product are evaluated as per HACCP guidelines to minimize the risk of potential health hazards among consumers. Baking is nowadays a more preferable technique of food processing rather than frying on grounds of long-term health benefits and retaining the natural taste and flavor of the finished product. Use of artificial sweeteners and leavening agents also impose long-term serious health risks to regular consumers by acting as diabetics.

The popular processing techniques in food sector are canning, fish processing, industrial rendering, tanneries, meat packing plants, slaughter houses, sugar industries and vegetable packaging plants.

Chapter 5

Food Irradiation

Irradiation leads to destruction and permanent damage to DNA of pathogenic organisms. As a result they lose their capability to multiply and proliferate. If insects are present in food items then they become sterile incapable of reproduction and plants lose their natural ripening processes.

The process of food irradiation is given the term 'cold pasteurization'. The energy density per atomic transition of ionizing radiation is very high. It is capable of breaking apart molecules and induce ionization, which is not achieved by simple heating. Ionizing radiations impart the same effect as of during heat pasteurization of liquids, such as milk.

Food irradiation is regarded as cold process of preservation because this kind of treatment does not cause any significant rise in temperature. The temperature of irradiated food product poses influence on physical changes induced by exposure to radiation. Rise in temperature induces increase in migration free radicals whish affect the overall rate of radiolysis. Reduced temperature decreases the production of volatile substances in food products. These volatiles are known to adversely affect the sensory quality of irradiated foods. Refrigeration on the other hand cause minimum level of such changes.

Nearly 40 countries all over the globe practice food irradiation and the process is permitted there. In an estimate, the volume of food treated is estimated to exceed 5,00,000 metric tonnes annually worldwide.

The effective irradiation of food depends on many factors such as, on the dose, some or all of the harmful bacteria and other pathogens present are killed[22]. Food irradiation also prolongs the shelf-life of food. In some foods such as herbs and spices food irradiation helps in reducing microbial count by several degree. In this way, spoilage causing microorganisms are permanently destroyed.

The U.S. Food and Drug Administration (FDA) has cleared among a number of other applications the treatment of hamburger patties to eliminate the residual risk of a contamination by a virulent *E. coli*. The United Nations Food and Agricultural Organization (FAO) have passed a motion to commit member states to implement irradiation technology for their national phyto-sanitary programmes

Including India, many countries like Australia, New Zealand, Thailand and Mexico have adopted under provision irradiation of fresh fruits for eradication of fruit fly. Countries like Brazil and Pakistan have also adopted the Codex Alimentarius Standard on Irradiated Food without any reservation or restriction.

Radiation Absorbed Dose

The measurement unit of exposure is Radiation absorbed dose which is the unit of physical quantity which regulates the processing of food products related to its beneficial effect.

Irradiation Dose Measuring Unit

The dose of radiation is measured in the SI unit known as the gray (Gy). One gray of radiation is equal to 1 joule of energy absorbed per kilogram of food material. In radiation processing of foods, the doses are generally measured in kilograys (kGy = 1,000 Gy).

Irradiation process does not find its wide acceptability among manufacturers due to negative influence on consumer satisfaction and perception. Many food producers also express their reluctance in this regard for the long term harmful effects. Even environmentalist activists and consumers believe that consumption of irradiated food expose to serious long term health hazards.

Other approved methods to reduce several pathogens in food include ultra-high temperature processing, UV radiation, ozone, heat-pasteurization, or fumigation with ethylene oxide.

Insect pests can also be eliminated by fumigating with aluminum phosphate, vapor heat, hot water dipping, or cold treatment, methyl bromide or forced hot air.

Other methods to extend shelf life of food items include freezing, flash freezing, modified atmosphere packaging, dehydration, carbon monoxide, vacuum packaging, including chemical additives.

The Food and Drug Administration (FDA) is the governing International agency for regulation of all aspects of food irradiation. It also emphasizes regarding use of radiations on suitable type of edible products, radiation dosing and proper labeling of radiated food products for consumers. The U.S. Department of Agriculture (USDA) is responsible for the inspection and monitoring of irradiated meat and poultry products and for the enforcement of FDA regulations concerning these irradiated food products. Since 1986, all irradiated products must carry the international symbol called a "Radura", which resembles a stylized flower.

FDA envisage on the proper presence and visibility of approved logo and declaration statement to appear properly on packaged irradiated foods like bulk containers containing unpackaged foods and/or on cardboard curtains containing such type of irradiated food items at the point of purchase, on invoices for irradiated ingredients and products to be sold to food processors.

Processors may add information explaining the need of irradiation for preventing spoilage or treated with irradiation instead of chemicals to control insect infestation.

Currently over 40 countries have approved irradiation as a process for preservation and increasing shelf-life of approximately 40 different foods. These include different fruits, vegetables, spices, grains, seafood, meat and poultry. More than half a million tonnes of food is now irradiated throughout the world on a yearly basis. The annual consumption of irradiated food items is constantly increasing.

According to agricultural researchers and scientists, implementation of organic farming techniques can help in reducing microbial load in food products. HACCP suggest that radiation measures up to specific limit can reduce pathogenic germs up to a great extent thereby rendering food safe for consumption.

Chapter 6

Biofermentation

Fermentation is a microbial technique and the reaction to be controlled in favorable and desirable conditions for food safety and quality after fermentation, especially in the production of alcoholic premium quality beverages like beer, wine and cider (Cavalieri *et al.*, 2003; Steinkraus, 1995; Ganguly, 2012a; 2013). The same technology is employed in the bred manufacturing industries for leavening activity brought about by the production of carbon dioxide by the microbial or yeast activity. The preservation effect during fermentation is attributed to the production of lactic acid in sour foods such as yoghurt, dry sausages, pickles, sauerkraut and vinegar (extremely diluted acetic acid) [Steinkraus, 1995; Harden and Young, 1906].

The term fermentation is derived from the Latin verb *fervere*, to boil, which describes the appearance of the action of yeast on extracts of fruit or malted grain during the production of alcoholic beverages. However, fermentation is interpreted differently by microbiologists and biochemists. Fermentation is a metabolic process converting sugar to acids, gases and/or alcohol. It occurs in yeast and bacteria, but also in oxygen-starved human muscle cells. In its strictest sense, fermentation is the absence of the electron transport chain and takes a reduced carbon source, such as glucose, and makes products like lactic acid or acetate. No oxidative phosphorylation is used, only substrate level phosphorylation, which yields a much lower amount of ATP. Fermentation is also used much more broadly to refer to the bulk growth of microorganisms on a growth medium. The science of fermentation is known as zymology.

The fermentation technology under controlled conditions is an age old practice both in households and industries for food processing and preservation, be it alcoholic beverage products of edible products derived from vegetable, fish and meat sources

(Steinkraus, 1995; Ruddle and Ishige, 2010). Louis Pasteur, the renowned French chemist is the world famous and first known zymologist in history, who in 1856 established the pivotal role of yeasts in fermentation. Pasteur originally defined fermentation as "respiration without air" after regular performances of lengthy experimental protocols (Dubos, 1995). After observation of the breakdown of sugars to alcohols by the action of yeast, the pioneer concluded that the entire reaction is driven by the chemical catalytic action of certain forces called ferments inside the yeast cells. It was further observed that the yeast extracts can bring bout fermentation of sugars even also in the absence of viable yeast cells. In 1897, Eduard Buchner of Humboldt University of Berlin, Germany discovered that sugars are fermented in the absence of viable cells also in the fermentation mixture. The yeast cells secrete a chemical component called zymase. For his memorable contributions in research and discovery of cell-free fermentation, in 1907 Buchner was awarded with the prestigious Nobel Prize in Chemistry. In 1906, NAD^+ was discovered out of studies carried out from ethanol fermentation (Lin *et al.*, 2012).

Fermentation is brought about by the conversion of sugars into ethanol chemically (Ganguly, 2012a). The fermentation technology applicable to food processing sector is also popularly known as zymology or zymurgy. Fermentation is an important and popular technique in food processing technology. It is resulted from the chemical reaction resulting from the breakdown of higher carbohydrates to alcohols and organic acids or alcoholic derivatives.

Microorganisms are capable of growing on a wide range of substrates and can produce a remarkable spectrum of products. The relatively recent advent of *in vitro* genetic manipulation has extended the range of products that may be produced by microorganisms and has provided new methods for increasing the yields of existing ones. The commercial exploitation of the biochemical diversity of microorganisms has resulted in the development of the fermentation industry and the techniques of genetic manipulation have given this well-established industry the opportunity to develop new processes and to improve existing ones.

The chemical equation below shows the alcoholic fermentation of glucose, whose chemical formula is $C_6H_{12}O_6$. One glucose molecule is converted into two ethanol molecules and two carbon dioxide molecules:

$$C_6H_{12}O_6 ' \rightarrow 2\,C_2H_5OH + 2\,CO_2$$

Before fermentation takes place, one glucose molecule is broken down into two pyruvate molecules. This is known as glycolysis.

Fermentation technology is primarily employed for the preservation of different food by production of acids and alcohols, biological fortification and enrichment of food items with potential biogenic products like essential amino acids, easily digestible proteins, essential fatty acids and useful vitamins, neutralization of anti-nutritional factors, to diversify and enrich the diet with various aromas, flavors and textures in food substrates and decrease in requirements of further processing techniques like cooking etc (Steinkraus, 1995).

Bio-fermentation technology makes it possible to grow a number of fungi in large tanks and in a matter of days large quantity can be produced. The technology has now become highly advanced and is making previously rare herbs like Cordyceps and Ganoderma much more accessible. Many studies indicate that the chemical nature of this biotechnology Cordyceps is almost identical to that of the wild variety. The Cordyceps contained in most commercial products is produced by this technology (Macfarlane and Macfarlane, 1993).

Advantages of Biofermentation

Standardization

Highly controlled by scientific means during its growth and "standardized", every batch is virtually identical and, in a sense, "perfect." This is very important in pharmaceutical terms because without standardization, it is difficult or impossible to develop drug-type standards for substances like Cordyceps. Once a herb can be standardized, all kinds of studies can be conducted that will be accepted by the scientific community. However, standardization, by itself, does not make a product effective.

Cost

Products can be produced more cheaply by bio-fermentation. High quality wild Cordyceps is rare. The culture-grown Cordyceps is available at a third of the cost, with approximately the same benefits. Fermentation technology makes this substance more readily available.

Vegetarian

Wild Cordyceps, by weight, is mostly caterpillar. For vegetarians, and those who do not want to consume caterpillars, the new technology provides a solution. The fermentation technology does not include the caterpillar in the growing process. The fungus is grown without the use of animal nutrients and the result is a 100% pure "vegetarian" health product (Macfarlane and Macfarlane, 1993).

Factors Regulating Fermentation

Physical factors

Temperature

Temperature has an impact on the growth and activity of different strains of yeast. At temperatures of 10 to 15°C, the non-*Saccharomyces* species have an increased tolerance to alcohol and therefore have the potential to contribute to the fermentation when the temperature increased, the maximum fermentation time gets shortened, but a much higher temperature inhibited the growth of cells and then the fermentation significantly declined (Lin *et al.*, 2012). This phenomenon may be explained because the higher temperature results in changing the transport activity or saturation level of soluble compounds and solvents in the cells, which might increase the accumulation of toxins including ethanol inside cells (Lin *et al.* 2012). Moreover, the indirect effect of high temperature might also be ascribed to the denaturation of ribosomes and enzymes and problems with the fluidity of membranes. However, at lower temperatures the cells showed lower specific growth rates which may be attributed to their low tolerance to ethanol at lower temperatures. It is commonly believed that 20 to 35ÚC is the ideal range for fermentation and at higher temperatures almost all fermentation would be problematic (Macfarlane and Macfarlane, 1993).

Influence of Substrate Concentration

The production of ethanol is affected by the substrate concentration higher substrate concentration may achieve higher ethanol production, but a longer incubation time was required for higher initial glucose concentrations. Moreover, higher initial glucose concentrations may have actually decreased the ethanol conversion efficiency when the pH value is not controlled, since the higher substrate and production concentrations may have inhibited the process of ethanol fermentation. More substrate did not improve the specific ethanol production rate when the pH value is not controlled.

Influence of pH

Improved ethanol fermentation activity can be achieved by controlling various parameters. In addition to temperature and substrate concentration, pH is also a key factor that affects ethanol fermentation. When the pH is lower than 4.0, the incubation time for maximum ethanol concentration became prolonged, but the maximum concentration is not very low. When the pH value is above 5.0, the quantity of ethanol produced substantially decreased. Therefore, a pH range of 4.0 to 5.0 may be regarded as the operational limit for the anaerobic ethanol production process (Lin *et al.*, 2012).

Chemical Factors

Several studies have been carried out to study the biochemical pathways followed during the degradation process of fish fermentation (Ruddle and Ishige, 2010). Strong odour in spoilt fish may be a reaction between TMAD and lactic acid producing TMA and acetic acid. The following are the chemical changes in deteriorating fish (Macfarlane and Macfarlane, 1993):

i.   enzymic degradation of nucleotides and nucleosides in the flesh leading to the formation of inosine, hypoxanthine, ribose, etc.

ii. bacterial reduction of trimethylamine oxide (TMAO), a non-volatile and non-odoriferous compound, to volatile trimethylamine (TMA) which has an amoniacal smell.

iii. formation of dimethylamine (DMA).

iv. breakdown of protein with subsequent formation of ammonia (NH_3) indole, hydrogen sulfide, etc.

v.   oxidative rancidity of the fat.

Volatile bases particularly TMA, DMA and NH, are associated with changes in the organoleptic and textural quality of fish (Ruddle and Ishige, 2010). The development of a specific aroma in fermented fish sauces and pastes may not be due to the action of micro-organisms. In a recent study it was found that micro-organisms play little or no part in aroma production. It can therefore be concluded that the microbiology of any salted, dried or fermented fishery product is greatly influenced by the natural micro-flora of the fish, the salt and the conditions under which processing takes place (Ruddle and Ishige, 2010).

There are certain risks and health hazards associated with excess and regular consumption of fermented food products. In Alaska, since 1985, there has been increase in incidences of botulism exceeding the case reported in the Americas. This is mainly caused for the practice of allowing whole fish, fish heads and meat of animals like sea lions, walrus, whale flippers, birds, seal tallow, beaver tails etc. to ferment for prolonged periods before consumption by the resident Eskimos there. During this extended fermentation, if plastic wrappers or containers are used, then *Clostridium botulinum* gets a conducive condition to thrive in the microaerophilic condition inside the plastic containers (Ganguly, 2012b; 2013).

Losses incurred at post-harvest are quite common and enormous leading to valuable food loss. At every stage of post-harvest practice, agricultural products are deprived from quality due to physical, chemical, biological and mechanical factors. In this article an overview has been presented on the major and common reasons for post-harvest food losses.

Food processing involves the conversion of raw ingredients into more acceptable food forms. Food processing is related to crops after harvesting, animal products prepared after slaughtering of animals and converting these products to appeal the general consumers for market profitability and for increasing the storage life of the finished processed products. Animal and fish feeds are also manufactured by this same mechanism of processing.

Fermentation is an important and popular technique in food processing technology. It is resulted from the chemical reaction resulting from the breakdown of higher carbohydrates to alcohols and organic acids or alcoholic derivatives. Fermentation is brought about by the conversion of sugars into ethanol chemically. The fermentation technology applicable to food processing sector is also popularly known as zymology or zymurgy.

Two types of radiations are used for preservation. They are ionizing radiations such as cathode and gamma rays and non-ionizing radiation such as infrared and UV rays. The latter two have poor penetration and bactericidal properties. Ionizing radiations act by disrupting progeny cycles of pathogenic microorganisms by destroying DNA/RNA cycles. Thus, it enhances shelf-life of food and reduces microbial contamination. Food processing by irradiation process destroys most of the pathogenic microorganisms. But is does not lead to sterilization of food. Consumers should also ensure proper storage of irradiated food materials under refrigeration followed by proper hygienic handling and cooking to reduce the risk of potentially harmful microorganisms. Irradiated foods maintain their wholesomeness and nutritive values. So, food irradiation increases the shelf life of food as well as its optimum quality is also maintained for longer durations. Even if some nutritional losses occur due to irradiation it is negligible.

Chapter 7
Need for Food Processing

As a result of increasing world population, the need for increased food supply has become an urgent and important consideration in many developing countries. Considerable efforts made in agricultural research and extension has resulted in increased crop yield resulting to increased food production.

Post-harvest losses of vegetables, fruits and fisheries are difficult to predict, the major agents producing deterioration mostly being attributed to microbiological causes and physiological damages. Post-harvest losses may be grouped broadly into food losses after harvesting and food losses due to social and economic reasons.

The losses at each stage of harvest and post-harvest practices due to improper handling can be large enough to result in a total loss of millions of food commodities every year. It is believed that a 50% reduction in post-harvest food loss in developing countries will reduce the need for food importation in these countries and will cause an increase in the food supply to meet the food demands. Also loss is far less than the amount of money that will be used to produce the same amount food.

However, it has been discovered that increased food production only is not the final solution if it is not complemented with adequate harvest and post-harvest practices. This is because good harvest and post-harvest practices will lead to reduction in the amount of food losses during and after harvest. The food moved from the farm, through the delivery system to the consumer must be presented in the good and acceptable from with little food loss during the movement. This is the ultimate goal of any food supply chain and not increased food production alone.

Global Acceptance

Nearly 40 countries all over the globe practice food irradiation and the process is permitted there. In an estimate, the volume of food treated is estimated to exceed 5,00,000 metric tonnes annually worldwide.

The effective irradiation of food depends on many factors such as, on the dose, some or all of the harmful bacteria and other pathogens present are killed. Food irradiation also prolongs the shelf-life of food. In some foods such as herbs and spices food irradiation helps in reducing microbial count by several degree. In this way, spoilage causing microorganisms are permanently destroyed.

The U.S. Food and Drug Administration (FDA) has cleared among a number of other applications the treatment of hamburger patties to eliminate the residual risk of a contamination by a virulent *E. coli*. The United Nations Food and Agricultural Organization (FAO) have passed a motion to commit member states to implement irradiation technology for their national phyto-sanitary programmes.

Including India, many countries like Australia, New Zealand, Thailand and Mexico have adopted under provision irradiation of fresh fruits for eradication of fruit fly. Countries like Brazil and Pakistan have also adopted the Codex Alimentarius Standard on Irradiated Food without any reservation or restriction.

Regulation of Global Food Irradiation

The Food and Drug Administration (FDA) is the governing International agency for regulation of all aspects of food irradiation. It also emphasizes regarding use of radiations on suitable type of edible products, radiation dosing and proper labeling of radiated food products for consumers. The U.S. Department of Agriculture (USDA) is responsible for the inspection and monitoring of irradiated meat and poultry products and for the enforcement of FDA regulations concerning these irradiated food products. Since 1986, all irradiated products must carry the international symbol called a "Radura", which resembles a stylized flower.

FDA envisage on the proper presence and visibility of approved logo and declaration statement to appear properly on packaged irradiated foods like bulk containers containing unpackaged foods and/or on cardboard curtains containing such type of irradiated food items at the point of purchase, on invoices for irradiated ingredients and products to be sold to food processors.

Processors may add information explaining the need of irradiation for preventing spoilage or treated with irradiation instead of chemicals to control insect infestation.

For preparing and formulating fish feed the economic aspect of each and every ingredient needs to get proper priority, as it is well known that appreciable amount of nutritional loss occurs during processing, heat treatment and storage. The nutritional requirements for fish mainly depends on the rate of growth of the fish in conjunct with additional influences like size, metabolic function in addition to the environmental influences and management strategies employed for rearing and breeding (Ganguly, 2013b).

Ideal aquafeed should contain adequate nutritional ingredients should be formulated as to contain all the essential components which should be balanced and adequate for the proper maintenance of growth, reproduction and overall health of the fishes (Ganguly, 2013b).

Diet for the fishes should be devoid of the harmful antinutritional factors (including mycotoxins majorly, aflatoxicosis, ochratoxicosis and zearalenone) which deteriorate the quality of the diet. The formulated diet should be well acceptable to the fishes and should not pose any adverse effect on their habitat or water system (Gatlin *et al.*, 1986).

A variety of polysaccharides from many sources behave as immunomodulators by stimulating the immune system. Increases in the interest in glucans as a result of experimental evidences have shown that 'zymosan' has the ability to stimulate macrophages by activating the complement system. They can be pharmacologically classified as biological response modifiers (BRM). Biological activities of α-1,3-glucans is influenced by different physicochemical parameters, such as solubility, primary structure, molecular weight, branching and polymer charge. During the development of immune reactions, immunomodulating effects of α-glucans are well established (Vetvicka and Sima, 2004). α-1,3-glucan possess a strong immunostimulating activity in a wide variety of species, including shrimps, fish (Ganguly *et al.*, 2009, 2010a), rats ,rabbits, guinea pigs, sheep, pigs , cattle, humans. Based on these results it has been concluded that α-1,3-glucans represent a type of immunostimulant that is active across the evolutionary spectrum. Invertebrates have active defense mechanisms which enable them to use their highly effective innate defense pathways against invading pathogens despite the absence of lymphocytes or antibody based adaptive immune system (Vetvika *et al.*, 2004; Ganguly et al., 2009, 2010a, 2010b, 2013a).

Nutritional Requirement at Different Stages of Growth

The nutrients in the aquafeed should be highly digestible to the fishes with high bioavailaibility. The feed should also have high storage life and losses due to physical and climatic factors should be minimum (Ganguly, 2013b).

The nutritional requirement among various fish types vary as per their habitat from freshwater to brackish water and to the marine system (Garling and Wilson, 1976). Fishes have varying nutritional requirement depending on their growth phase *i.e.* from larval, fishling, spawning and up to table fish stage. During the period of maximum growth, the requirement for potential nutrients also rises (Cowey, 1975).

Role of Optimum Availability of Proximate Principles

The availability of minerals varies among various fish species and sources. Phosphorus digestibility in some feeds by channel catfish or rainbow trout is much higher than by the stomachless carp.

The fatty acid content also varies linearly with it. The vitamin and mineral composition should be in conjunct with the major proximate principles present in the aquafeed. The energy presence in the diet varies according to the size of the fish species as it is met up by the presence of carbohydrate content (Cho and Kaushik, 1985).

Under normal processing and storage conditions the amino acids, several vitamins and inorganic nutrients are relatively stable to heat, moisture and oxidation. Some of the vitamins are recommended for use in excess of the requirement, as they are subject to some loss. On the contrary, excess fortification with vitamins and micronutrients beyond permissible limits may lead to some loss in the nutrient content majorly during feed processing (Ganguly, 2013b).

The protein incorporation in the formulated diets should be in optimum ratio with the energy component present therein. Majorly, the protein component should consist of all the essential amino acids (Cho and Kaushik, 1985). The technical or reagent grade compounds fetch more mineral sources and are more consistent than from usual feedstuffs (Goldstein and Forster, 1970). In the diet formulation for fishes, the protein finds the upper hand as the most important proximate component in the diet. The overall digestion coefficient of the diet depends on the availability of superior and qualitative protein source.

Chapter 8

Feed Additives for Immunomodulation

Biological activities of α-1,3-glucans is influenced by different physicochemical parameters, such as solubility, primary structure, molecular weight, branching and polymer charge. During the development of immune reactions, immunomodulating effects of α-glucans are well established (Vetvicka and Sima, 2004). α-1,3-glucan possess a strong immunostimulating activity in a wide variety of species, including shrimps, fish, rats, rabbits, guinea pigs, sheep, pigs, cattle, humans. Based on these results it has been concluded that α-1,3-glucans represent a type of immunostimulant that is active across the evolutionary spectrum. A variety of polysaccharides from many sources behave as immunomodulators by stimulating the immune system. Increases in the interest in glucans as a result of experimental evidences have shown that 'zymosan' has the ability to stimulate macrophages by activating the complement system. They can be pharmacologically classified as biological response modifiers (BRM). Invertebrates have active defense mechanisms which enable them to use their highly effective innate defense pathways against invading pathogens despite the absence of lymphocytes or antibody based adaptive immune system (Vetvika *et al.*, 2004; Ganguly *et al.*, 2009, 2010a, 2010b, 2013a).

α-1,3-glucans are usually isolated from cell walls of bacteria, mushrooms, algae, cereal grains, yeasts and fungi and are structurally complex homopolymers of glucose (Zekovic and Kwiatowski, 2005).

Anti-microbial immune mechanisms found in invertebrates can be induced by fungal α-glucans (Brown and Gordon, 2005). Protease cascades initiated by PAMP induce majority of recognition of these responses in the haemolymph. Cascade activation results in coagulation in the haemolymph, or anti-microbial peptide secretion by immunocompetent cells. α-glucan recognition can also result to phagocytosis by certain haemocytes . α-glucans are recognized by fishes as foreign agents because of their similarity to fungal or bacterial Gram negative polysaccharides. An inflammatory response is produced as a result by the immune system of fishes after exposure to provide effective protection. (Robertsen *et al.*, 1994).

Innate immunity plays a very important role in combating microbial infection in all animals. Surface determinants conserved among microbes are recognized by receptors which activate the innate immune response, but absent in lipopolysaccharides, peptidoglycans and mannans (Medzhitov and Janeway, 1997, Ganguly *et al.*, 2009, Ganguly *et al.*, 2010b). These receptors upon recognition activate multiple and complex signalling cascades leading to regulation of transcription of target genes encoding effector molecules. Specific transcription programmes elicited by different pathogens can be investigated by using microarray technology (De Gregorio *et al.*, 2001). Diseases possess a major problem in aquaculture production, especially for the invertebrate farming (Ganguly *et al.*, 2009, 2010a, 2010b, 2013a).

Under intensive conditions, fishes are more susceptible to microbial infections especially in their larval stages. During stress, immunostimulants can provide resistance to pathogens. Only a few immunostimulants can be used in aquaculture. Glucans are commercially significant as immunostimulating agents. Different types of α-glucans have been used successfully to increase resistance of fish and crustaceans against bacterial and viral infections (Cook *et al.*, 2003; Bagni *et al.*, 2005). It has been seen that health, growth and general performance of many different animal groups, including farmed shrimp, fish and land animals may be improved by the use of α-glucans. Product source, animal species, development stage of the target organism, dose and type of glucan, route and time schedule of administration affect the immunomodulatory effects of glucans (Guselle *et al.*, 2007; Ganguly et al., 2009, 2010a, 2010b, 2013a) and the association with other immunostimulants. Many studies have been carried out to measure the effects of glucan on fish immunity. Some investigators have adopted the *in vitro* culture of macrophages with glucan (Cook *et al.*, 2003), but *in vivo* studies have been carried out by majority of the workers (Sahoo and Mukherjee, 2001).

The skin, respiratory tract and digestive tract of fishes is an open system constantly contacting with a surrounding environment and water. Undoubtedly, the microflora of the environment plays an important role in the formation of the microflora of the digestive tract of fishes (Strom and Olafsen 1990; Hansen *et al.* 1992). Compared to water, digestive tract is an ecosystem far richer in nutrients and therefore more favorable for the growth of the majority of bacteria. Definitely, not all bacteria in food which gain entry in the digestive tract of fishes establish themselves there. Part of them adapts themselves in the digestive tract, whereas the others are digested by the enzymes produced by the host organism. Gastrointestinal microorganisms feed on the food of the host organism, which is digested by the enzymes produced by them and by the latter. As a result, 'chymous' is formed, the composition of which decides the abundance and qualitative composition of communities of gastrointestinal microorganisms.

In the digestive tract, bacteria that feed on materials produced by other microorganisms during metabolism are found too. Several competitive reviews on the microflora of the digestive tract of fish in the last decades of the 20th century have convincingly proved that the digestive tract of marine and fresh-water fish has a regular microflora, which could be divided into autochthonous (regular normal microflora) and allochthonous (Syvokieni 1989; Ringo and Birkbeck, 1999). It is found that populations of microorganisms are far more abundant in the digestive tract of fish than in water, but as far as the diversity of the qualitative composition of bacterioflora is concerned the digestive tract of hydrobionts is poorer compared to that of aquatic microflora (Sakata, 1990).

Bacteriocenoses of the digestive tract of hydrobionts are poorer compared to those of homoeothermic animals. In the digestive tract of homoiothermal animals obligatory anaerobic bacteria predominate (Finegold *et al.* 1983), whereas in that of hydrobionts and aerobic and facultative anaerobic do so, though in the digestive tract of some fish obligatory anaerobic bacteria too are found (Sakata and Yuki 1992). It was proved that the microflora of the digestive tract of fish plays an important role in the formation of resistance to infectious diseases, for it produces antibacterial materials preventing pathogenic bacteria from getting into an organism (Sugita *et al.* 1998). Gastrointestinal bacteria take part in the decomposition of nutrients and provide the host with physiologically active materials, such as enzymes, amino acids and vitamins (Sugita *et al.* 1997). The aim of this work was the exploration of beneficial effect of gut bacteria in the provision of active nutrient principles to the host (fish). The innovation was discovered later as the active role played by microflora in imparting resistance to fish against infectious diseases.

Chapter 9
Nature of Microflora in Fishes

Microflora Present on Fish Skin

Aeromonas hydrophila, *Pseudomonas* and *Vibrio* are the bacteria present on the skin of fishes. This bacterial population is generally influenced by the marine ecosystem. The bacterial population along with the slimy coating on the scales of the fish body provides an efficient barrier against the entry of virulent microorganisms through skin of fishes. Bacteria associated with the fish skin can be enumerated by acridine orange epifluorescence microscopy and by plate counts on several media.

Microflora Present in Fish Respiratory System

Fishes breathe through their gills. The interesting thing in fish is a long bony cover for the gill that can be used for pushing water. Some fishes pump water using the operculum. When they swim, water flows into the mouth and across the gills. Freshwater fish use a type of countercurrent flow to maximize the intake of oxygen that diffuse through the gill. Countercurrent flow occurs when deoxygenated blood moves through the gill in one direction while oxygenated water moves through the gill in the opposite direction. This mechanism maintains the concentration gradient thus increasing the efficiency of the respiration process as well. So, the only types of microflora evidenced in fish respiratory system are symbionts which generally occur in their surrounding aquatic environment. Till date, no pathogenic microflora or microflora of aquatic interest regarding fish respiratory system have been reported.

Microflora residing on skin has no role to play in nutrient digestion and metabolism in fish.

Microflora Present in Fish Digestive Tract

It is known that the structure of the digestive tract in different fish species differs. Differences are seen already in the early stages of fish development. So, the first factor influencing the formation of gastrointestinal bacteria communities is the structure of the digestive tract (Lesel *et al.* 1986). The formation of the regular microflora in the digestive tract of fish larvae and fry is a complex process and depends on fish spawn (lay and deposit large quantities of eggs in water) [Scott 1997; Meisner and Burns 1997] food and the microflora of the surrounding water. After studying the formation of the microflora of the digestive tract of carp from the larval stage to adult fish it has been proved that in the digestive tract of fish the bacterioflora is formed gradually (Vesta Skrodenyte-Arbaeiauskiene 2000). The *Aeromonas, Pseudomonas, Clostridium* and *Bacteroides* bacteria predominate. In the digestive tract of fish, bacteria of the genus *Bacteroides* appear as late as on the 44th day after hatching (Cahill, 1990). Later they become predominant in the intestines of adult fish. Investigation results suggest that bacteria of the genera *Aeromonas, Pseudomonas* and *Flavobacterium-Cytophaga* prevail in the bacteriocenoses of the digestive tract of freshwater fish (Ringo and Birkbeck 1999; Vesta Skrodenyte-Arbaeiauskiene 2000; Voverienë *et al.* 2002). The impact of feeding intensity and food on the qualitative and quantitative composition of intestinal bacteriocenoses of fish was studied by a number of scientists (Ringo and Olsen 1999; Pucci *et al.* 2004; Vesta Skrodenyte-Arbaeiauskiene 2000; Voverienë et al. 2002). The structure of intestinal bacteriocenoses of fish is influenced by farming conditions of fish, too (Ringo and Strom, 1994). *Aeromonas* and *Lactobacillus* bacteria prevail in the intestinal bacteriocenoses of fish inhabiting natural water bodies, whereas *Enterobacteriaceae*, which may make up to 50% of all bacteria, are prevalent in the bacteriocenosis of fish raised in farms and fed on artificial food. Microflora of the digestive tract of fish is investigated intensively and in different aspects, but data about the impact of xenobiotics on the intestinal microflora of hydrobionts are few. If bacteria non typical of the living environment of hydrobionts are abundant in the water surrounding the latter, they make a negative influence on the immune system of fish by restraining it, thus impacting the animal's general physiological state (Cahill, 1990).

Role of Microflora in Starch and Cellulose Digestion and Metabolism

Starch is a major ingredient of feed for freshwater fishes such as carps and Tilapia (Takeuchi 1991). The starch ingested is hydrolyzed into its constituent sugars and oligosaccharides in the digestive tract of fishes. The enzyme amylase is widely present in the digestive tract of freshwater fishes and plays an important role in the digestion of starch. Enzyme found in intestinal lumen of fishes could have potentially come from either the pancreas or the secretary cells of gut wall. In addition, enzymes from intestinal microflora have a potential role in digestion, especially for substrates such as cellulose which few animals can digest and other substrates as well (Smith 1989). The intestinal tract of fishes is generally colonized by great number of heterophilic bacteria including aerobes and anaerobes. Many ecological studies on gut microflora of fish have been presented from time to time (Cahill, 1990).

Fish in general utilize dietary carbohydrate poorly. Furthermore, different types of carbohydrates may not be equally available to fish (Furuichi and Yone 1982). Dietary fibre generally refers to all indigestible plant matter mainly cellulose and other complex polysaccharides. It has long been considered an inert and insignificant part of an animal's diet, mainly because it was believed to contribute little nutritionally. Vegetable biomass is abundant in many freshwater environments and cellulose, being the main structural material of plants, is the most abundant carbohydrate in nature (Lindsay and Harris 1980). However, little work has been carried out to determine the ability of predominantly herbivorous fish to utilize the structural polysaccharides which constitute a major element of the total calorific value of plant material. Until the demonstration by Davies (1965) of a significant cellulolytic bacterial population in the gut of a number of species of non-ruminant animals, it had generally been accepted that nutritional consequential cellulolytic activity was confined to ruminant species. Dietary utilization of cellulose in fishes occurs due to presence of specific enzymes which may be either endogenous or exogenous. Indian major carps, which are poikilothermic animals, produce cellulase endogenously and there a stable cellulolytic microflora present in their digestive tract Cellulolytic bacteria present in fish digestive tract has a influence in metabolism and there exists a correlation between the degree of cellulolytic activity and feeding habit.

Immunomodulatory Substances Secreted by Microflora of Digestive Tract of Fish

The intestinal bacteria produce some bioactive substances like tetrodotoxin (Noguchi *et al*. 1987), eicosa-pentanoic acid (Yazawa et al. 1988), biotin (Sugita et al. 1992), Vitamin B_{12} (Sugita et al. 1991) and antibacterial substances (Westerdahl et al. 1991) which may benefit the host fish. There exists strongly a symbiotic relationship between fish with its constituent microflora.

Effect of Microflora of Fish System in Biodegradation

In the environment, constant processes of bio-destruction of materials occur carried out by various organisms. Some species of microorganisms easily utilize metabolites produced by other microorganisms and thus develop intensively, whereas other species develop through using organic and mineral pollutants. Hydrocarbon biodegradation carried out by populations of natural microorganisms is the major mechanism removing petroleum and other hydrocarbons from the environment (Leahy and Colwell, 1990). Based on the number of published reports, the most important hydrocarbon-degrading bacteria both in marine and soil environments are *Achromobacter, Acinetobacter, Alcaligenes, Arthrobacter, Bacillus, Flavobacterium, Nocardia, Pseudomonas* and the *coryneforms* (Leahy and Colwell, 1990).

Aquaculture is one of the fastest growing industries in the world. There is a need for enhanced disease resistance, feed efficiency and growth performance of cultured fish species. The cost of production are likely to be reduced if growth performance and feed efficiency are increased in commercial aquaculture, Also if the survival of the fishes increase, then their overall production cost would be remarkably reduced. Dietary supplementation of different feed additives *e.g.* immunostimulants, probiotics and prebiotics usually in small quantities for the purpose of fortifying it with certain nutrients have been found to be beneficial for improving immune status, feed efficiency and growth performance of crustaceans and finfishes.

Chapter 10
Immunostimulants, Probiotics and Prebiotics

Immunostimulants are substances that stimulate the immune system by inducing activation or increasing activity of any of its components. The term immunostimulant can be used interchangeably with immunomodulator, adjuvant and biological response modifier. Immunostimulators are drugs and nutrients which stimulate particularly cells of the monocyte/ macrophage system and thereby modulate and potentiates the immune system of the body.

There are two main categories of immunostimulators:

a) **Specific immunostimulators** are those which provide antigenic specificity in immune response, such as vaccines or any antigen. For specific immune response host should have prior exposure to an antigen so that recognition and subsequent activation occurs through co-ordinated effort of B-lymphocytes and T-cells.

b) **Non-specific immunostimulators** are those which act irrespective of antigenic specificity to augment immune response of other antigen or stimulate components of the immune system without antigenic specificity, such as glucans, synthetic drug levamisole etc. Many endogenous substances are non-specific immunostimulators. For example, Glucans and mannans possess non-specific immunostimulatory effect. α-glucan is a polymer of glucose consisting of linear backbone of α-1,3 linked D-glucopyranosyl residues with varying degree of branching from the C_6 position (Bohn and Bemiller, 1995). α-glucans are major components of yeasts, mushrooms and fungal mycelia. Mannan is a plant polysaccharide that is a polymer of the sugar mannose. Detection of mannan leads to lysis in the mannan binding lectin pathway.

Immunostimulants as a feed additive, significantly provides protection against pathogen and upregulates phagocytosis, bacterial killing and oxidative burst.

Mechanism of Ativation of Immune System

Anti-microbial immune mechanisms found in invertebrates can be induced by fungal â-glucans (Brown and Gordon, 2005). α-glucans are recognized by fishes as foreign agents because of their similarity to fungal or bacterial Gram negative polysaccharides. An inflammatory response is produced as a result by the immune system of fishes after exposure to provide effective protection. (Robertsen *et al.*, 1994).

Application of Immunostimulants in Aquaculture

Diseases possess a major problem in aquaculture production, especially for the invertebrate farming (Bachère, 2003). Glucans having strong immunomodulating activity has been well studied in fishes (Anderson, 1992). Only a few immunostimulants can be used in aquaculture (Siwicki *et al.*, 1998). Many studies have been carried out to measure the effects of glucan on fish immunity. Some investigators have adopted the *in vitro* culture of macrophages with glucan (Cook et al., 2001), but in vivo studies have been carried out by majority of the workers (Sahoo and Mukherjee, 2001; Ortuno et al., 2002). Under intensive conditions, fishes are more susceptible to microbial infections especially in their larval stages (Smith et al., 2003). During stress, immunostimulants can provide resistance to pathogens. Glucans are commercially significant as immunostimulating agents. Different types of â-glucans have been used successfully to increase resistance of fish and crustaceans against bacterial and viral infections (Paulsen et al., 2001; Bagni et al., 2005). It has been seen that health, growth and general performance of farmed shrimp and fish may be improved by the use of α-glucans. Product source, animal species, development stage of the target organism, dose and type of glucan, route, time schedule of administration and the association with other immunostimulants affect the immunomodulatory effects of glucans (Guselle *et al.*, 2007).

The immunostimulatory effects of glucan, chitin, lactoferrin, levamisole, vitamins B and C, growth hormone and prolactin have been reported in fish and shrimp. These immunostimulants mainly facilitate the function of phagocytic cells and increase their bactericidal activities. Several immunostimulants also stimulate the natural killer cells, complement, lysozyme and antibody responses of fish. The most effective method of administration of immunostimulants to fish is by injection. Oral and immersion methods have also been reported, but the efficacy of these methods decreases with long-term administration. Overdoses of several immunostimulants induce immunosuppression in fish. Growth promoting activity has been noted in fish or shrimp treated with glucan or lactoferrin. Immunostimulants can overcome immune suppression by sex hormones.

For the effective use of immunostimulants, the timing, dosages, method of administration and the physiological condition of fish need to be taken into consideration. Immunostimulants can reduce the losses caused by disease in aquaculture; however, they may not be effective against all diseases.

Probiotics

The term "probiotics" was coined by Parker (1974) to describe "organisms and substances which contribute to intestinal microbial balance". They affect the host animal by improving its intestinal microbial balance (Fuller, 1989). Probiotics are viable cultures of bacteria and fungi which when introduced through host feed have a positive effect on host health. Some reside in the digestive tracts of the individuals while others have an external origin. They are also sometimes referred to as 'Direct Fed Microbials (DFM)'. Probiotics can be used as growth promoters and also for therapeutic purposes.

In the gut, a variety of population of microorganisms are present and their population is affected by various factors namely, age, diet, environment, stress and medication.

The most commonly used organisms in probiotic preparations are lactobacilli, streptococci and bifidobacteria. Except these, *Bacillus* spp., yeasts, *Saccharomyces* spp. and filamentous fungi (*Aspergillus oryzae*) are also used as probiotics. The probiotic preparations are available as tablets, powders, capsules, pastes or sprays.

Characteristics of Probiotics

The good probiotics should possess the following characteristics:

i) It should be resistant to pH and bile acids

ii) It should be non-pathogenic.

iii) It should possess a high viability.

iv) It should be stable on storage and in the field.

v) It should survive the gut environment and should have the potentiality to colonize in gut.

vi) It should be cultivable on a large scale.

vii) It should have the ability to attach to the gut epithelial lining.

viii) It should provide a beneficiary effect to the host animal.

Even if any new strains are used for probiotic development then they should possess all these aforementioned characteristics.

Mode of Augmentation of Immune System

The probiotic microorganisms in the gut stimulate the immune response of host system in two ways. They can migrate through the gut wall as viable cells thereby multiplying to a limited extent or the antigens which are being released by the dead organisms can be absorbed and stimulate the immune response directly. Probiotics generally find their applications in aquafeed because of their effects on high growth rate, improved feed conversion and improved resistance to diseases. Probiotics show an positive effect on host immune response through increased activity of macrophages shown by enhanced ability to phagocytose organisms or carbon particles, increased production of systemic antibody *e.g.* IgM and interferon and increased effect of local antibody at mucosal surfaces such as gut wall. The effect of probiotics on the host immune system can be measured by estimating the levels of macrophage enzymes.

Application of Probiotics in Aquaculture

Lactic acid bacteria have received priority as probiotics in fish feed (Hagi *et al.*, 2004). Lactic acid bacteria produce acetate and lactate which proves helpful for inhibiting the growth of several species of Vibrio (Vazquez et al., 2005). Lactic acid bacteria when included in diet of Atlantic cod increased their survivility rate when they were challenged by pathogen *Vibrio anguillarum*. Use of probiotics influence the specific and non-specific immunity in many fish species like rainbow trout (Nikoskelainen *et al.*, 2003; Panigrahi *et al.*, 2005) and gilthead seabream (Salinas *et al.*, 2005). Probiotics help in reducing the mortality of larval stages of different fishes and pathogen-challenged fishes and they provide the needed enzymes useful for digestion. But, effectiveness of these probiotics is adversely affected by harsh conditions of extrusion or pellet manufacturing.

Extrusion technique is a process in food processing technology which combines several unit operations including mixing, cooking, kneading, shearing, shaping and forming. Food extrusion is a form of extrusion used in food processing. It is a process by which a set of mixed ingredients are forced through an opening in a perforated plate or die with a design specific to the food, and is then cut to a specified size by blades. The machine which forces the mix through the die is an extruder, and the mix is known as the extrudate. The extruder consists of a large, rotating screw tightly fitting within a stationary barrel, at the end of which is the die. Extrusion cooking is a high-temperature short-time (HTST) process which reduces microbial contamination and inactivates enzymes. The main method of preservation of both hot- and cold-extruded foods is by the low water activity of the product (0.1–0.4), and for semi-moist products in particular, by the packaging materials that are used.

The principles of operation in extrusion are similar in all types: raw materials are fed into the extruder barrel and the screw(s) then convey the food along it. Further down the barrel, smaller flights restrict the volume and increase the resistance to movement of the food. As a result, it fills the barrel and the spaces between the screw flights and becomes compressed.

As it moves further along the barrel, the screw kneads the material into a semi-solid, plasticized mass. If the food is heated above 100°C the process is known as *extrusion cooking* (or *hot extrusion*). Here, frictional heat and any additional heating that is used cause the temperature to rise rapidly. The food is then passed to the section of the barrel having the smallest flights, where pressure and shearing is further increased. Finally, it is forced through one or more restricted openings (dies) at the discharge end of the barrel as the food emerges under pressure from the die, it expands to the final shape and cools rapidly as moisture is flashed off as steam. A variety of shapes, including rods, spheres, doughnuts, tubes, strips, squirls or shells can be formed. Typical products include a wide variety of low density, expanded snack foods and ready-to-eat (RTE) puffed cereals.

Cold extrusion, in which the temperature of the food remains at ambient is used to mix and shape foods such as pasta and meat products. Low pressure extrusion, at temperatures below 100°C, is used to produce, for example, liquorice, fish pastes, surimi and pet foods.

Physical Properties

Expansion ratio (ER)

When moisture content of the feeding material increases, there is decrease in the specific mechanical energy (SME), apparent viscosity, and radial ER during extrusion of maize grits. Parsons *et al.* (1996) reported a decrease in the ER of corn meal when the extrusion moisture content was increased from 19.5 to 21.5% (w/w). Kokini et al. (1992) and Della Valle *et al.* (1997) explained a sharp decrease in volumetric expansion with increased moisture content by the shrinkage and collapse of the extrudate after maximum expansion. Most studies recognize that gelatinized starch plays a major role in expansion by providing the gas-holding capacity to the extrudate melt, whereas other ingredients such as proteins, sugars, fats, and fiber act as diluents or dispersed phase fillers that reduce the stretchability of the starchy matrix. Conway (1971) reported that the lower limit of starch content for good expansion is 60–70%.

Water hydration (WH)

WH capacity increased with extrusion temperature and, in general, at any specific extrusion temperature WH decreased with increased moisture content. Higher WH might result from a greater extent of starch gelatinization (McPherson *et al.*, 2000). WH also is greatly affected by the degree of porosity or expansion of the extrudate, as higher porosity and thinner cell walls in the extrudates lead to greater water absorption.

Hardness

The hardness of cereals increased with moisture content for each extrusion temperature. In general, the hardness of extruded cereal exhibited an inverse relationship with extrudate expansion, as observed in several studies on extruded products where hardness was represented by instrumentally measured mechanical properties such as compression modulus and crushing stress. Hardness is greatly affected by the expansion of the extrudates.

Aguilar-Palazuelos et al. (2006) conducted a study to analyze the effects of extrusion barrel temperature (75–140°C) and feed moisture (16–30%) on the production of third-generation snacks expanded by microwave heating. A blend of potato starch (50%), quality protein maize (QPM) (35%), and soybean meal (SM) (15%) was used in the preparation of the snacks. A laboratory single extruder with a

1.5×20.0×100 mm die-nozzle and a central composite routable experimental design were used. Expansion index (EI) and bulk density (BD) were measured in expanded pellets, viscosity at 83°C (V83), thermal properties, and relative crystallinity were measured in extruded pellets. EI increased and BD decreased when the barrel temperature was increased, while the feed moisture effect was not significant. V83 increased when feed moisture increased. Extrusion modified the crystalline structures of the pellets and the X-ray data suggests the formation of new structures, probably due to the development of amylose-lipid complexes. The maximum expansion of pellets was found at barrel temperatures of 123–140°C, and feed moisture of 24.5–30%. It is possible to obtain a functional third-generation snack with good expansion characteristics using a microwave oven, and this snack has health benefits due to the addition of QPM and SM.

Aguilar-Palazuelos et al. (2006) concluded that the mathematical model used in analyzing the data from the extrusion study was satisfactory for the evaluated responses, with values of R2e"0.77, lack-of-fite"0.131, CVd"9.01% (except V83) and P of F (model) < 0.005. The barrel temperature was the variable that most affected the expansion index (EI) and bulk density (BD) and the feed moisture had a significant effect on the V83. Increasing barrel temperature and decreasing feed moisture, probably favored the degradation of starch in extruded products as shown by DSC, viscoamylograph properties, and X-ray diffraction analysis. Response surface methodology showed the best expansion of third-generation extruded products at 28% feed moisture and 130°C barrel temperature. In addition, the products obtained in this processing zone were probably not completely degraded. Therefore, it is possible to produce third-generation snacks using extrusion technology that have a significant nutritional and nutraceutical value by using high-protein quality maize (QPM) and soybean meal (SM).

Chemical Properties

Lipid oxidation is the major chemical challenge for preservation of food. This oxidation can reduce the nutritive quality by decreasing the content of essential fatty acids, such as linolenic acid (C18:3) and linoleic acid (C18:2), which are essential fatty acids. These long-chained unsaturated fatty acids are highly susceptible to oxidation. High temperature of extrusion can increase the pro-oxidant transition metal concentration, particularly iron, due to the metal wear on extruder parts. Neutral, inorganic form of minerals, e.g. iron, has been reported to promote oxidation.

Currently, extrusion-cooking as a method is used for the manufacture of many foodstuffs, ranging from the simplest expanded snacks to highly-processed meat analogues. The most popular extrusion-cooked products include:

- Direct extruded snacks, RTE (ready-to-eat) cereal flakes and a variety of breakfast foods produced from cereal material and differing in shape, color and taste and easy to handle in terms of production;

- Snack pellets – half products destined for fried or hot air expanded snacks, precooked pasta;

- Baby food, pre-cooked flours, instant concentrates, functional components;
- Pet food, aquafeed, feed concentrates and calf-milk replacers;
- Texturized vegetable protein (mainly from soybeans, though not always) used in the production of meat analogues;
- Crisp bread, bread crumbs, emulsions and pastes;
- Baro-thermally processed products for the pharmaceutical, chemical, paper and brewing industry;
- Confectionery: different kinds of sweets, chewing gum.

There are also many regulatory issues regarding the application of probiotics in aquafeed. Growth phases of the animal, the type of dosing used and health status of animal have also an effect. Sometimes desired outcomes are not obtained after use of probiotics in feed. This is attributed to the fact that different probiotics contain different microorganisms which may act differently under variable situations and also they have their own metabolic pathways which differ from the others.

Prebiotics

Prebiotics are the indigestible components present in the diet which are metabolized by specific microorganisms which prove to be helpful for growth and health of the host (Manning and Gibson, 2004). Certain nutrients such as, linoleic acid, linolenic acid, and soluble carbohydrate were studied for their effects on the aerobic/facultative anaerobic intestinal microbiota of Arctic char Salvelinus alpinus (Ringo and Olsen, 1999). Prebiotics shift the microbial community to one dominated by beneficial bacteria, such as *Lactobacillus* spp. and *Bifidobacterium* spp. (Manning and Gibson, 2004). Till date, no significant data is available related to the use of prebiotics in fish feed. In an experiment, when linoleic acid was supplemented to the diet of Arctic char the total viable counts increased by an order of magnitude (10-fold) as compared with fish fed a diet without linoleic acid (Ringo, 1993). Polyunsaturated fatty acids of the n-3 and n-6 series also were shown to alter the microbial population of Arctic char, with the lactic acid bacteria Carnobacterium spp. being the dominant facultative anaerobe cultivated (RingØ et al., 1998).

The effects of a potential prebiotic has also been investigated on hybrid striped bass *Morone chrysops* X *M. saxatilis* (Li and Gatlin, 2004, 2005) by application of 'GroBiotic®-A' which is a mixture of partially autolyzed brewers yeast, dairy ingredient components, and dried fermentation products. It was also found to be encouraging that fish fed a diet containing 'GroBiotic®-A' had a significantly higher feed efficiency and significantly lower mortality when challenged with the bacterial pathogens Streptococcus iniue and Mycobacterium murinum.

Mechanism of Action on Immune System

Prebiotics have the potentiality to enhance many host biological responses and they also reduce the mortality of fishes due to invasion by pathogens. Prebiotics have the potential to enhance numerous biological responses while lowering mortality due to microbial pathogens. However, the anaerobic intestinal tract microbiota of commercially important fishes, such as channel catfish, hybrid striped bass, tilapia and salmonids need to be investigated to determine if there are any particular bacterial species to be enhanced with the use of prebiotics. By increasing the production of VFAs in the GI tract, the host will be benefited by recovering some of the lost energy from indigestible dietary constituents and by inhibiting potential pathogenic bacteria (Manning and Gibson, 2004; Vazquez *et al.*, 2005). The VFAs produced are also indicative of the microbial population present in the GI tract (Nisbet, 2002).

Herbivorous fishes such as sea chubs *Kyphosus cornelii* and *K. sydneyanus* were the first species shown to have VFAs as bacterial metabolic by-products in their intestinal tracts (Choat and Clements, 1998). Other fishes that have been found with bacterial VFAs in their intestinal tracts include tilapia *Oreochromis mossambicus* (Titus and Ahern, 1988). Prebiotics have many beneficial effects such as increased disease resistance and improved nutrient availability. Prebiotics have much potential to increase the efficiency and sustainability of aquaculture production.

The most commonly used prebiotic preparations in aquaculture are fructooligosaccharide (FOS), transgalactooligosaccharide (TOS), inulin, glucooligosaccharide, xylooligosaccharide, isomaltooligosaccharide, soybeanoligosaccharide, polydextrose, lactosucrose (Vulevic *et al.*, 2004; Propulla, 2008). Natural sources of prebiotics in vertebrates include chicory, onion, garlic, leek, tomato, honey etc.

Properties of Prebiotics

Prebiotics should have the following properties:

i)   It should be easy to incorporate in the feed or ration.

ii)   It should regulate viscosity of gut.

iii) It should be non-carcinogenic

iv)   It should be derived from dietary polysaccharides.

v)   It should have low calorific value.

vi)   It should reduces harmful microbial load.

vii) It should be effective at lower concentration.

viii) It should exert anti-adhesive properties against harmful gut microbes.

ix)   It should stimulate the beneficial gut microbes.

x)   It should not possess any residual effects.

Popular Prebiotics

The current most popular targets for prebiotics use are lactobacilli and bifidobacteria largely based on their success in the concerned area. Prebiotics can be used as unique tools to create gut microflora with controlled composition which can be correlated with specific physiological conditions.

Prebiotics like organic acids etc. are mainly used in order to sanitize the feed containing various infectious and pathogenic agents (Hinton *et al.*, 1985; Barchieri and Barrow, 1996; Thompson and Hilton, 1997). Organic acids (OA) in their undissociated forms are able to pass through the cell membrane of the bacteria, where they dissociate to produce H^+ ions which lower the pH of bacterial cell causing the organism to use its energy to restore the normal balance. Whereas the RCOO⁻ ions, produced from the acid can disrupt DNA, hampering protein synthesis and putting the organism in stress. As a result the organism cannot multiply rapidly (Nursey, 1997).

Prebiotics are non-digestible feed ingredients that beneficially affect the host by selectively stimulating the growth or activity of one or a limited number of bacterial species, already resident in the gut and thus attempt to improve host health (Gibson and Roberfroid, 1995). Mainly prebiotics are small fragments of carbohydrates and commercially available as oligosaccharides of galactose, fructose or mannose.

Among these, mannan oligosaccharide (MOS) obtained from *Saccharomyces* spp. of yeast outer cell wall maintain gut health by adsorption of pathogenic bacteria, containing type-I fimbriae or by agglutinating different bacterial strains (Spring *et al.*, 2000).

A variety of polysaccharides from many sources behave as immunomodulators by stimulating the immune system. Increases in the interest in glucans as a result of experimental evidences have shown that 'zymosan' has the ability to stimulate macrophages by activating the complement system. They can be pharmacologically classified as biological response modifiers (BRM). Biological activities of α-1,3-glucans is influenced by different physico-chemical parameters, such as solubility, primary structure, molecular weight, branching and polymer charge. During the development of immune reactions, immunomodulating effects of α-glucans are well established (Vetvicka and Sima, 2004). α-1,3-glucan possess a strong immunostimulating activity in a wide variety of species, including shrimps, fish, rats ,rabbits, guinea pigs, sheep, pigs , cattle, humans. Based on these results it has been concluded that α-1,3-glucans represent a type of immunostimulant that is active across the evolutionary spectrum. Invertebrates have active defense mechanisms which enable them to use their highly effective innate defense pathways against invading pathogens despite the absence of lymphocytes or antibody based adaptive immune system (Vetvika *et al.*, 2007; Ganguly *et al.*, 2009, 2010). α-glucans can be used as environmental friendly agents, in contrast to the chemical anti-microbial products in aquaculture which is frequently affected with disease problems under intensive conditions. Nowadays, the demand for health promoters, higher feed efficiency and alternatives to antibiotics has increased in aquaculture. α-glucans possess probiotic effects and have immunomodulating activities. Many studies are been done which focus on the properties of these glucans and their efficient use in fish farming (Ganguly *et al.*, 2009, 2010).

Xylanase is the name given to a class of enzymes which degrades the linear polysaccharide beta-1, 4 xylan to xylose, thus breaking down hemicelluloses which are a major component of the cell wall of the plants. Xylanases are known to increase protein digestibility of wheat and this is attributed particularly to release of protein from the xylan enriched aleurone layer. Xylanase supplementation improves conjugated bile acid function in intestinal contents and increase villus size of small intestine wall (Bar *et al.*, 2012). The addition of Xylanase improves weight gain, feed intake, feed efficiency, AME and decreased water intake (Wu *et al.*, 2004) and Vitamin E content of liver was significantly improved by addition of xylanase (Danicke *et al.*, 1999, 2001). Nutri-xylanase is a bacterial xylanase processed from *Bacillus subtilis* and produced by a micro-filtration advanced fermentation technique.

Chapter 11

Effect of Nutritional Non-Antibiotic growth Promoters on Live Body Weight Gain

Eidelsburger and Kirchgessner (1994) reported that calcium formate alone or in combination with other acids when given at the rate of 0.5% and 1.5%, increased FCR and growth performance up to 35 days of age. Benedetto (2003) also observed mix of organic acids (ACIDLAC) used as a replacer of growth promoters (AGPs) and improved production performance along with other beneficial effects. Mairoka *et al*. (2004) also reported that mixture of organic acids can be effectively used as a substitution of antibiotic growth promoters (AGPs) for improved physiological performance. Savage *et al*. (1997) concluded from a dose responsive study that MOS @ 0.11%, maximized weight gain up to 0-8 weeks of age. Stanley *et al*. (2000) found same type of effect with supplementation of 0.1% MOS on body weight gain. Parks *et al*. (2001) reported from a study with MOS that MOS may be utilized as an alternative to AGPs to improve turkey performance.

Evidence of the beneficial effects of probiotics gave rise to the concept of prebiotics, which selectively stimulate the growth of and/or activate the metabolism of one or a limited number of health-promoting bacteria in the gut, thus improving the host's intestinal balance. Information pertaining to application of prebiotics in aquaculture is extremely limited to date. Research has shown that dietary supplementation with a commercial prebiotic significantly enhanced growth and disease resistance of hybrid striped bass beyond that achieved with brewers' yeast (Gatlin and Li, 2004, Li and Gatlin, 2004).

It has already been reported that 1% formic acid or 1.45% calcium formate have no effect on the live weight (Izat *et al.*, 1990). It was found out that 80% formic acid and 20% propionic acid mixture added at 1% level in ration did not affect live weight (Kaniawati *et al.*, 1992). It has also been reported that formic acid and propionic acid mixture (85% and 15%) added at 1% level to the ration in the initial period did not affect weight gain (Visek, 1978). Reports have also been made about significant increase in body weight gain with the supplementation of 0.5% lactic acid in drinking water (Veeramani *et al.*, 2003). It was also revealed that increase in body weight takes place with supplementation of lactic acid.

Effect of Nutritional Non-antibiotic Growth Promoters on Dressing Percentage and Weight of Vital Organs and Muscles

Pelicano (2005) reported that there was effect on the dressing percentage and weight of different organs and muscles at 21 and 42 days without any major influence on the dressing percentage, organ and muscle weight under different treatment groups with organic acid salts individually and its combination. Higher villus height in duodenum, jejunum in small intestine was reported with most organic acidifier in diet (Loddi *et al.*, 2004). Again it was reported higher villus height in the ileum with the diet based on organic acidifier compared with diet fed without MOS + organic acidifier (Savage *et al.*, 1997). Therefore, the supplementation of organic acidifier may increase villus height of different parts of small intestine. So organic acidifier reduces the growth of many pathogenic and non-pathogenic intestinal bacteria, decreases intestinal colonization and reduces infections process, ultimately decreasing inflammatory process at the intestinal mucosa. It increases villus height and function of secretion, digestion and absorption of nutrients can be appropriately performed by the mucosa (Iji and Tivey, 1998). It was also reported positive effects of the use of prebiotics on the intestinal mucosa among which a significant increase in villus height of three segments of small intestine of birds, age one week and supplemented with MOS (Maiorka *et al.*, 2004). The use of organic acidifier in diet significantly increased villus height of different segment of small intestine possibly by reducing intestinal colonization of pathogenic and non-pathogenic bacteria (Savage *et al.*, 1997, Loddi *et al.*, 2004) respectively.

The studies have determined the effect of Fermacto prebiotic addition in feeds, on the growth and food conversion ratio in common carp fry. Four types of granulated experimental feeds were prepared, from which three feeds contained different amounts of Fermacto prebiotic (F1 – 1.0 g/kg, F2 – 2.0 g/kg, F3 – 3.0 g/kg) and a control feed K – without a prebiotic. During a 50-day growth test, the fish receiving prebiotic feeds showed significantly higher mean individual body weight (p d" 0.05) in comparison with the control group. The best effects were obtained in the variant F3, where the specific growth rate (SGR) was 2.44%, feed conversion ratio (FCR) was 1.21, protein efficiency ratio (PER) was 2.18, and the values significantly differed from the remaining variants. The values of the coefficients of feed protein retention were contained in the interval: 2.80-4.53%, while the values of fat retention coefficients ranged from 8.97 to 15.07%. During the growth test, no fish losses were recorded. Feeding of carp fry with feeds containing an addition of Fermacto prebiotic improves the rearing results; the optimal addition of prebiotic is 3 g of the preparation per 1 kg of feed (Mazurkiewicz *et al.*, 2008).

Effect of Nutritional Non-antibiotic Growth Promoters on Increase in Feed intake and Digestive Performance

Loddi (2003) described higher villi in the intestinal mucosa (duodenum) of birds fed with MOS at 7 and 21 days of age respectively. Pelicano *et al.* (2005) reported that in jejunum MOS + OA resulted in higher villi in the jejunum (p<0.01) followed by the diets containing MOS based prebiotics. While in case of ileum the higher villi length was observed when birds were fed with prebiotic based on MOS, compared to the control group.

Microorganisms that is sensitive to acid pH and results in higher villi length (Radecki and Yokoyama, 1991). Some bacteria may recognize binding sites on the prebiotics instead of intestinal mucosa and the colonization by pathogenic bacteria in intestine is thus reduced. Therefore, besides a lower infection incidence, there is an increase in the absorption of available nutrients, a mechanism that directly affects the recovery of the intestinal mucosa, increasing villi length. These results disagree to those obtained by Pelicano *et al.* (2003) and Santin *et al.* (2001) respectively, who found no difference in ileal villi length with the use of probiotics and prebiotics.

Effect of Nutritional Non-antibiotic Growth Promoters on Gut Microbial Load

It is reported that MOS and OA significantly reduce bacterial load in the intestine (Newman, 1994, Lon, 1995, Spring *et al.*, 2000, Fairchild *et al.*, 2001). Stanley *et al.* (2004) concluded that yeast cell culture residue (YCR) treatment resulted in lower intestinal coliform population in comparison to control and other antibiotic treated (lasalocid @ 90.7kg/ton, bacitracin @50gm/ton) groups. Sims *et al.* (2004) also found that MOS + BMD treatment resulted in significantly lower *Clostridium perfringens* population in the gut. MOS is believed to block type-I fimbriae and prevents pathogens from attaching to the intestinal lining and passes them out of the gut. (Dawson and Pirulescu, 1999). For this reason, MOS treated birds showed lens microbial load in the gut. On the other hand organic acids in their undissociated forms pass through the cell membrane of the bacteria, and dissociate to produce H^+ ions which lower the pH of the bacterial cell, causing the organism to use its energy, trying to restore the normal balance. Whereas $RCOO^-$ anions produced from the acid can disrupt DNA, hampering protein synthesis and putting the organism in stress. As a result the organisms cannot multiply rapidly and decreases in number (Nursey, 1997).

Chapter 12

Herbal Feed Supplements

A herbal immunomodulator is a substance which stimulates or suppresses the components of immune system including both innate and adaptive immune responses (Agarwal and Singh, 1969). The modulation of immune system by various medicinal plant products has become a subject for scientific investigations currently worldwide.

Under Indian scenario, poultry industry has become a means for earning livelihood for the economically distressed farmers in India due to its promising results in productivity and National economy. Poultry rearing is currently the fastest growing industry in our National livestock sector which is benefiting us from production and advantages in prices along with provision of proteinacous food.

Mode of action in Immunostimulation of Different Herbal extracts

Many herbal plant preparations are prescribed to strengthen host resistance (Thatte and Dahanukar 1986). Many useful plants fall under this category. They exhibit immunomodulatory activities. One such plant, *Tinospora cordifolia*, commonly called 'Guduchi' has been examined for its immunomodulatory properties. Guduchi means to rejuvenate dead cells. It is widely used in veterinary folk medicine and has also been claimed to be beneficial according to 'Ayurveda' for the cure of jaundice, skin diseases, diabetes, anemia, emaciations and various infections for its anti-spasmodic, anti-inflammatory, anti-arthritic and anti-allergic properties (Chopra *et al.* 1982). It has also been reported that it improves the phagocytic and bactericidal activities in patients suffering from polymorphism in surgical jaundice (Thatte *at al.* 1989). Kolte *et al.* (2007) studied the effect of feeding *T. cordifolia* in broiler birds which were immunosuppressed with cyclophosphamide. They had found a significant rise in antibody titer against ND virus with augmentation of inflammatory reaction to skin contact sensitivity test. Rege *et al.* (1989) and Bishavi *et al.* (2002) have proved the hepato-protective effect of *T. cordifolia*. Manjrekar *et al.* (1999) also found that aqueous extract of *T. cordifolia* is capable of increasing leukocyte count in mice.

Also, *Ocimum sanctum*, commonly known as *'tulsi'* is also used in Ayurveda for various ailments including treatment of allergies. The plant has been reported to evince significant anti-stress properties. The beneficial effects of *O. sanctum* could therefore be due to its direct or indirect effect on the immune system. *O. sanctum* has been reported to modulate humoral immune response by releasing mediators for hypersensitivity reactions (Kujur 2001; Krishnamohan *et al.* 1997; Kumar 2003).

Withania somnifera also fall in this category with many other useful plants. They exhibit immunomodulatory activities. *Withania somnifera* (commonly called 'Ashwagandha') root extracts possess anti-estrogenic, adaptogenic, anti-cancer and anabolic activities having beneficial effects in the treatment of arthritis, geriatric problems and stress. The root of *Asparagus racemosus* (commonly called 'Satavar') possess anti-diarrheal, anti-ulcerative, anti-spasmodic, aphrodisiac, galactogogue and other properties and has therefore gained its importance in Ayurveda, Siddha and Unani systems of medicine (Nadkarni, 1954). It has been observed that feeding *W. somnifera* and *A. racemosus* dried root powder significantly stimulates both humoral and cell mediated immune responses in swiss albino mice by Kuttan and Kuttan (1992). *W. somnifera* and *A. racemosus* extracts increase phagocytic activities of macrophages in vitro (Rege and Dahanukar 1993). There have been studies on the immunomodulatory activities of *W. somnifera* and *A. racemosus* in mice with myelo-suppression induced by cyclophosphamide, azathioprim or prednisolone. Extracts of *W. somnifera* and *A. racemosus* have also shown immunopotentiating effects in cyclophasphamide treated mouse with ascitic sarcoma (Diwanay *et al.* 2004). Kalita and Dutta (1999) reported that maternal antibody was persistently found in sera samples tested against ND virus during the first week of age in broilers. This was attributed to transfer of natural passive immunity in young chicks as demonstrated by Hellar (1975). Muruganandan *et al.* (2001) reported the effects of ethanolic extracts of *W. somnifera* and *A. racemosus* on humoral immune system which was assessed by humoral immune response and cell mediated immune response in mice.

The use of various plant extracts and herbal fed additives in a specific dose during the scheduled vaccination regimen may be helpful in obtaining higher protective antibody against different infections including production and development of more effective cell mediate immune response for protection against various bacterial, viral and other diseases. Herbal formulation may be therefore recommended for use as positive immunomodulator in normal and immunocompromized susceptible animals and birds.

The importance of various herbal ingredients and plant derivatives in immunomodulation is a subject for scientific investigations currently worldwide. The mystery of *Moringa oleifera* as a vegetable source for immunomodulation in human and its capacity to impart protection against diseases by building host resistance has made it an important and economical nutritional supplement majorly in developing countries. The coconut fruit obtained from the coconut palm has numerous medical and commercial benefits. The various health related properties of coconut water, coconut milk, coconut cream, creamed coconut and other derivatives from the fruit have been highlighted in this article. The abstract provides an overall summary of different utilities and profile of different products obtained from the coconut for which it is being used as a principal ingredient by almost every consumer in Asian and Western countries in cooking and eating practices. The various derivatives from the fruit are well acceptable to consumers of all age groups, having certain limitations its high saturated fat content. Coconut has been recently proved to be a source of saturated fat that would not elevate the lipid profile in the body, except High Density Lipoprotein (HDL), which is good for health and absolutely no contraindications now to any age. Coconut is a highly valued ingredient in our eating practice for its enormous medical benefits. However, due to its high lipid and saturated fat content it is discouraged in the diet of patients suffering from cardiovascular ailments and hypertension. The major importance of the fruit is valued for the great medicinal properties of coconut water and the flesh of the fruit. The meat of mature coconut is a flavoring and texture improving ingredient in Indian and Asian homemade food. Lime is having tremendous importance in our diet and regular feeding habits due to its enormous nutritional and natural medicinal benefits from ancient time in India. It keeps away many diseases, improves resistance to diseases and provides long term profits by purifying blood in the body system naturally. The present review has been constructed considering the future scope of research in immunomodulation in medical and veterinary sciences which can be explored from the different vegetative resources available naturally to us.

Many herbal plant preparations are prescribed to strengthen host resistance.[1,2] Many useful plants fall under this category. They exhibit immunomodulatory activities. One such plant, *Tinospora cordifolia*, commonly called 'Guduchi' has been examined for its immunomodulatory properties. Guduchi means to rejuvenate dead cells. It is widely used in veterinary folk medicine and has also been claimed to be beneficial according to 'Ayurveda' for the cure of jaundice, skin diseases, diabetes, anemia, emaciations and various infections for its anti-spasmodic, anti-inflammatory, anti-arthritic and anti-allergic properties.[3] It has also been reported that it improves the phagocytic and bactericidal activities in patients suffering from polymorphism in surgical jaundice.[2] Kolte et al.[4] studied the effect of feeding *T. cordifolia* in broiler birds which were immunosuppressed with cyclophosphamide. They had found a significant rise in antibody titer against ND virus with augmentation of inflammatory reaction to skin contact sensitivity test. Rege et al. and Bishavi et al. have proved the hepato-protective effect of *T. cordifolia*. Manjrekar et al. also found that aqueous extract of *T. cordifolia* is capable of increasing leukocyte count in mice.

Also, *Ocimum sanctum*, commonly known as '*tulsi*' is also used in Ayurveda for various ailments including treatment of allergies. The plant has been reported to evince significant anti-stress properties. The beneficial effects of *O. sanctum* could therefore be due to its direct or indirect effect on the immune system. *O. sanctum* has been reported to modulate humoral immune response by releasing mediators for hypersensitivity reactions.

Withania somnifera also fall in this category with many other useful plants. They exhibit immunomodulatory activities. *Withania somnifera* (commonly called 'Ashwagandha') root extracts possess anti-estrogenic, adaptogenic, anti-cancer and anabolic activities having beneficial effects in the treatment of arthritis, geriatric problems and stress. The root of *Asparagus racemosus* (commonly called 'Satavar') possess anti-diarrheal, anti-ulcerative, anti-spasmodic, aphrodisiac, galactogogue and other properties and has therefore gained its importance in Ayurveda, Siddha and Unani systems of medicine. It has been observed that feeding *W. somnifera* and *A. racemosus* dried root powder significantly stimulates both humoral and cell mediated immune responses in swiss albino mice by Kuttan and Kuttan[13]. *W. somnifera* and *A. racemosus* extracts increase phagocytic activities of macrophages *in vitro*.

Moringa oleifera is a highly valued plant, distributed in many countries of the tropics and subtropics. Moringa is nature's medicine cabinet. It is best known as excellent source of nutrition and a natural energy booster. Different parts of this plant are being employed for the treatment of different ailments in the indigenous system of medicine. The plant has tremendous pharmacological action and pharmaceutical application too. It possesses analgesic, anti-inflammatory, antipyretic, anti-asthmatic and wound healing properties. Also, it possesses anti-diabetic, anti-cancerous and hepatoprotective properties too.

Extract from the seeds is used as a flocculant in a low cost form of water treatment. It effectively helps in bacterial reduction in edible water. The seeds are also considered an excellent biofuel source for making biodiesel.

Medicinal Properties of Moringa Oleifera

Moringa leaves and pods are helpful in increasing breast milk in the breastfeeding months. One tablespoon of leaf powder provides 14% of the protein, 40% of the calcium, 23% of the iron and most of the vitamin A needs of 1-3 years aged children. Six tablespoons of leaf powder will provide nearly all of the woman's daily iron and calcium needs during pregnancy and breastfeeding. The Moringa seeds yield 38–40% edible oil (called ben oil from the high concentration of behenic acid contained in the oil). The refined oil is clear and odorless and resists rancidity at least as well as any other botanical oil. The seed cake remaining after oil extraction may be used as a fertilizer or as a flocculent to purify water. The bark, sap, roots, leaves, seeds, oil and flowers are used in traditional medicine in several countries. The Moringa tree has great use medicinally both as preventative and treatment. Much of the evidence is anecdotal as there has been little actual scientific research done to support these claims. India's ancient tradition of ayurveda says the leaves of the Moringa tree prevent 300 diseases. There has been reports on significant antibiotic activity of this tree.

Medicinal Value of Coconut Water

Coconut water is considered to be sterile unless the fruit is damaged from an external source. There have been reports of coconut water used for intravenous administration where normal saline solution for medical purpose was unavailable in developing countries or on the war front. Coconut water is rich in mineral content with high potassium and anti-oxidant contents which has various medical utilities. Coconut water also contains cytokinin which is one of the beneficial components in it. Coconuts in which water to be used for drinking purpose are harvested from the coconut palms when they appear green in color. Coconuts sometimes due to natural calamities fell on the ground and they are susceptible to get damaged and get exposed for being damaged by insects or pests and animals. Coconut water acts as a natural energy or sports drink, as it is rich in mineral content especially in potassium levels. Coconut water has a high demand among consumers for its nil fat content and low contents of carbohydrates, calories, and sodium. Coconut water serves as a potential healthy drink for adults and old persons as it has promising health utilities. Coconut milk has tremendous importance especially in Ayurvedic traditional medicinal purposes. It is generally used to maintain the electrolyte balance and to rule out dehydration losses. Also, it is used for treatment of ulcers in the mouth. Some recent studies have suggested that coconut milk has anti-microbial properties in the gastrointestinal tract, hyperlipidemic balancing qualities and useful for topical applications. In addition, the coconut milk contains auric acid as saturated fat which has medicinal utilities in the cardiovascular system.

Medicinal Utility of Lime

Limes are acidic in nature and serve as rich source of vitamin C, citric acid, sugar, certain minerals like calcium and phosphorus. Fresh lime juice possesses medicinal property which is well known from ancient ages in India. It is also called as sacred fruit in the *Vedas*. Sharangdhara and Charaka. The latter two famous physicians of ancient India had used the lime juice for alleviating orthopedic aliments therapeutically. The vitamin C as a primary component of the lime juice increases the resistance of individuals to several diseases, helps in wound healing and increases the health of eyes. It improves the maintenance of good dentition and keeps away toothache, dental caries, swollen gums, fragility of bones and bleeding of the gums. Lime is vital in the treatment of gastric disorders like indigestion, constipation and peptic ulcer. It stops the occurrence of indigestion, burning in the chest due to high acidity in the stomach, abrupt bilious vomiting and excessive accumulation of saliva in the mouth. Lime juice with a pinch of soda bicarb mixed in water improves indigestion and gastric upset due to severe acidity after heavy rich/ spicy meals. It can also be used as carminative in indigestion and sedation is produced by release of carbonic acid and gas. In chronic constipation by promoting biliary secretion from the liver, it improves intestinal motility. The acidic juice of lime facilitates the lipid and alcohol absorption and neutralizes excessive bile produced by the liver. The juice reduces gastric acidity by counteracting with the effects of greasy food. It is, therefore, useful in the treatment of peptic ulcers. Acidic juice of the fruit acts as curative for tonsillitis. Oral ingestion of lime juice mixed with salt in water provides relief from burning sensation and also stops bleeding in cystitis (inflammation of urinary bladder). It is also a recommended therapy in weight reduction and in obesity.

During mornings at empty stomach, lime juice with honey in lukewarm water to be ingested for 2-3 months for effective weight reduction. Of course, low calorie diet is also a must along with this.

Importance of Other Miscellaneous Herbs and Plants

The plant derived and herbal growth promoters supplemented in the diet or added in the drinking water in the broiler and poultry birds have a promising biological effect on their growth performance, to reduce the pathogenic bacteriological load in different parts of digestive tract and to increase villus height in different segments of small intestine mainly in duodenum. The plant derived growth promoter enhance productive performance of the broiler in terms of body weight gain with minimum alteration of gut morphology and the possibility of bacterial invasion can be regulated.

It summarizes the wide use of various plant extracts and herbal fed additives in a specific dose during the scheduled vaccination regimen may be helpful in obtaining higher protective antibody against different infections including production and development of more effective cell mediate immune response for protection against various bacterial, viral and other diseases. Herbal formulation may be therefore recommended for use as positive immunomodulator in normal and immunocompromized susceptible animals and birds.

The neem bark, leaves, and seeds are used to make medicine. Less frequently, the root, flower, and fruit are also used. Salimuzzaman Siddiqui was the first scientist to discover the antibacterial, antiviral, anthelmintic and antifungal, constituents of the Neem tree. In 1942, he extracted three bitter compounds from neem oil, which he named as nimbin, nimbinin, and nimbidin respectively. *Centella asiatica*, commonly known as centella, is a small, herbaceous, annual plant of the family *Mackinlayaceae*, or subfamily *Mackinlayoideae* of family *Apiaceae*, and is native to Asian countries. It is commonly used in in Ayurvedic medicine, traditional African medicine, and traditional Chinese medicine as a medicinal herb. For its potential application as herbal medicine for therapeutic purposes, the present review attempts to highlight on the various research based facts and issues related to it. The papaya fruit is very common in Asian continent for its wide culinary and medicinal uses. The raw papaya is cooked as vegetable, while the ripe one is served in salads as fruit. It has a characteristic aroma and deliciously sweet and tender to taste. The present article discusses on its potentiality of various medicinal herbs and fruits for use in folk and traditional Asian and African medicine as it are a reserve source of dietary enzymes and vitamins.

Neem leaf is used diseases of the heart and blood vessels (cardiovascular disease), fever, diabetes, gum disease (gingivitis), for leprosy, eye disorders, bloody nose, intestinal worms, stomach upset, loss of appetite, skin eczema, ulcers and hepatic disorders. Papaya is also applied topically for the treatment of cuts, rashes, stings and burns (Ganguly, 2014*a*). Papain, a notable protease remains present in papaya. It is believed that it can raise platelet levels in blood. Papaya may be used as a medicine for malaria and dengue fever for its antimalarial and antiplasmodial properties. The raw papaya and its leaf are also used for meat tenderizing for the papain content in the fruit (Ganguly and Bordoloi, 2014). Centella is a plant which grows in tropical swampy areas. The stems of Centella are slender with creeping stolons, green to reddish-green in color, connecting plants to each other. It has long-stalked, green, reniform leaves with rounded apices which have smooth texture with palmately netted veins. The leaves are borne on pericladial petioles nearly 2 cm length. Centella grows in low lying wet areas along ditches. As the plant is aquatic in habitat, it is especially sensitive to pollutants in the water (Ganguly and Bordoloi, 2014).

Multivarious Biomedical Implications and Importance in Ayurvedic System of Medicine

Centella helps in the increase of hemoglobin in blood. It also promotes improvement in the venous system and encourages positive connective tissue growth. The herb is recommended for the treatment of various skin conditions such as leprosy, lupus, varicose ulcers, eczema, psoriasis, diarrhea, fever, amenorrhea, diseases of the female genitourinary tract and also for relieving anxiety and improving cognition. *Centella asiatica* is a popular addition to many skin creams and therapies. Extract of centella can be found in body slimming tonics, body-firming products, and anti-aging products. For decades, this herb has proven itself to be miraculous in terms of scar and wound healing. *Centella asiatica* is useful in alleviating many maladies, including stress, asthma, hemorrhoids and even leprosy. Doctors have used it successfully in pregnant women during and post pregnancy. It can be used to discourage varicose veins and stretch marks, and then again after birth for healing (Ganguly and Bordoloi, 2014).

The neem bark is used for malaria, stomach and intestinal ulcers, skin diseases, pain, and fever. The flower is used for reducing bile, controlling phlegm, and treating intestinal worms. The neem fruit is used for treatment of urinary tract disorders, bloody nose, phlegm, eye disorders, hemorrhoids, intestinal worms, diabetes, wounds, and leprosy. Neem twigs are used for cough, asthma, hemorrhoids, intestinal worms, low sperm levels, urinary disorders, and diabetes. The seed and seed oil are used for leprosy and intestinal worms. They are also used for birth control and to cause abortions. The stem, root bark, and fruit are used as a tonic and astringent. Some people apply neem directly to the skin as a skin softener and in treatment of skin diseases, wounds, and skin ulcers and as a mosquito repellent. Neem contains chemicals that might help reduce blood sugar levels, heal ulcers in the digestive tract, prevent conception, kill bacteria and prevent plaque formation in the mouth. People in the Indian villages and towns practice the chewing of neem twigs instead of using toothbrushes. Neem twigs are often contaminated with fungi within 2 weeks of harvest and should be avoided. Neem leaves are burnt to repel insects and flies from the crop fields (Ganguly, 2014a).

Neem products are popular and predominantly in demand in Ayurveda for its antibacterial, antiviral, contraceptive, anthelmintic, antifungal, antidiabetic, and sedative properties. In Ayurvedic and Unani medicine it is recommended for skin diseases. Neem oil is also used for detoxification of blood, to balance blood sugar levels, healthy hair and to improve liver function. Neem leaves have been also been used to treat skin diseases like eczema, psoriasis, etc, (Ganguly, 2014*a*).

Papaya fruit is a source of nutrients such as provitamin A carotenoids, vitamin C, folate and dietary fiber. Papaya skin, pulp and seeds also contain a variety of phytochemicals, including lycopene and polyphenols (Ganguly , 2014*b*).

Significance For use in Cosmetics and Topical Supplements

Papaya is a good topical supplement for treatment of acne, skin infections and wounds. The flesh of papaya is rich in dietary fiber and thus helps in lowering blood cholesterol. It prevents premature ageing and in treatment of endoparasitic worms in gastrointestinal tract. Papaya fruit aids in proper digestion and prevents constipation. It keeps colon infection away and helps in curing morning sickness and nausea. In addition, it contains anti-inflammatory enzymes which help in curing osteoporosis in elderly people and in alleviating joint arthritis (Ganguly, 2014*b*).

In obese people, it acts as a weight reducer and in controlling body weight. Papaya helps in preventing menstrual cramps and helps in maintaining regular menstrual flow. Papaya is a rich source of Vitamins A and C and thus promotes immunity (Ganguly, 2014*b*).

Shampoos and soaps contain papaya extracts as it helps in preventing dandruff and hair fall. It helps in regulating the growth of cancerous cells in hepatic cancers and also prevents renal problems by inducing antioxidant and oxidative free radical scavenging (Ganguly, 2014*b*).

Chapter 13
Biological Growth Promoters

A variety of pathogenic organisms may gain access into milk and milk products from different sources and cause different types of milk-borne intoxications in susceptible human beings. Milk and milk products may carry organisms as such or their toxic metabolites. Ingestion of preformed toxins in milk and milk food products produces poisoning syndromes in the consumers. So, adequate preventive and control measures should be implemented for production of hygienic and edible dairy products.

Exotoxins from bacteria alter the appearance, odor and flavor of food. Food-borne illness can be prevented by proper cooking, adequate heat treatment and refrigeration of food products. Certain guidelines have been laid down by the Food Safety and Inspection Service of USDA as essential in preventing bacterial food-borne illness.

α-glucans can be used as environmental friendly agents, in contrast to the chemical anti-microbial products in aquaculture which is frequently affected with disease problems under intensive conditions. Now-a-days, the demand for health promoters, higher feed efficiency and alternatives to antibiotics has increased in aquaculture. α-glucans possess probiotic effects and have immunomodulating activities. Many studies are been done which focus on the properties of these glucans and their efficient use in fish farming (Ganguly *et al.*, 2009, 2010a, 2010b, 2013a).

Preparation of proper and optimum dietary formulation for fishes is a primary prerequisite for sustainable and profitable aquaculture, especially in consideration of the long term economic benefit for the rural fisherfolk community (Ganguly, 2013b).

Immunostimulants, probiotics and prebiotics have been investigated and reported to have numerous beneficial effects in aquaculture such as increased disease resistance and improved nutrient availability which provides increased sustainability and profitability of finfish and crustacean production. So, more research regarding usage and effect of different immunostimulants, probiotics and prebiotics is warranted.

The enzyme producing microbes isolated from fish digestive tracts can be beneficially applied as probiotics with the feed containing high amount of cellulosic and starchy materials. The main strategy in the use of probiotics is to isolate intestinal bacteria with favorable properties from mature animals and include large numbers of these bacteria in the feed of immature animals of the same species (Gildberg *et al.* 1997). The use of probiotics has a long tradition in animal husbandry (Stavric and Kornegay, 1995), but is rarely applied in aquaculture. The use of probiotics in commercial aquaculture is necessary for formulating diets at larval stages to minimize the cost of feed preparation. The enzyme producing microorganisms isolated from the fish digestive tracts in the present study can be beneficially used as a probiotic while formulating the diet for fish, especially in the larval stages. There is need for enhanced survivability, disease resistance, feed efficiency and growth performance. Dietary supplementation of different feed additives *e.g.* immunostimulants, probiotics and prebiotics usually in small quantities have been found to improve immunity, feed efficiency and growth performance of crustaceans and fishes (Ganguly *et al.* 2010).

There is need for enhanced survivability, disease resistance, feed efficiency and growth performance. Dietary supplementation of different feed additives e.g. immunostimulants, probiotics and prebiotics usually in small quantities have been found to improve immunity, feed efficiency and growth performance of crustaceans and fishes (Ganguly *et al.*, 2012). So, organic acid salts and MOS can be used as alternatives to growth promoters but their combination strategy can be used to achieve good health and growth performance. It can also be commented that supplementation of enzymes on enhances productive performance in terms of body weight gain.

However, more sufficient researches are in demand to be carried out to establish the medicinal facts of the mentioned indigenous plants and fruits. In infants, use of neem oil is fatal. Some disadvantages of neem includes miscarriages, abortions and infertility. Although, American Cancer Society recommends the centella herb for having anticancerous properties, but research in this regard are in progress. Preliminary experimental animal researches have proved that the papaya seeds have potential contraceptive and abortifacient effect, but is non-teratogenic for the presence of phytochemicals in it.

The dietary prebiotics *viz.*, dietary organic acid (OA) supplements, mannan oligosaccharide (MOS), α-glucan and xylanase supplementation are mainly used in order to enhance live body weight gain, dressing percentage, weight of vital organs and muscles and mean villus lengths in digestive tract of poultry birds. Prebiotics can also act as immunostimulants. The term immunostimulant can be used interchangeably with immunomodulator, adjuvant and biological response modifier. Immunostimulators can be in the form of drugs and nutrients. They stimulate the monocyte-macrophage system and thereby modulate the immune system of the body.

Prebiotics are non-digestible feed ingredients that beneficially affect the host by selectively stimulating the growth or activity of one or a limited number of bacterial species, already resident in colon and thus attempt to improve host health (Gibson and Roberfroid, 1995; Ganguly and Mukhopadhayay, 2011). Mainly prebiotics are small fragments of carbohydrates and commercially available as oligosaccharides of galactose, fructose or mannose (Ganguly, 2013a,b).

Among these, mannan oligosaccharide obtained from *Saccharomyces* spp. of yeast outer cell wall maintain gut health by adsorption of pathogenic bacteria, containing type-I fimbriae or by agglutinating different bacterial strains (Spring *et al.*, 2000) and increase villi length uniformity and integrity (Loddi *et al.*, 2004). Effects of buffered propionic acid in presence and absence of bacitracin or roxarsone were reported earlier (Izat *et al.*, 1990) in which significant increase in dressing percentage for female broilers and a significant reduction in abdominal fat of males at 49 days (Versteegh and Jongbloed, 1999) tested the effect of dietary lactic acid on performance of broilers from 0 to 6 weeks age. Body weight gain tended to be greater, whereas feed to gain rations were significantly improved where birds were fed 2% lactic acid as prebiotic. Beneficial effects of different organic acids like formic acid, propionic acid, lactic acid ammonium formate and calcium propionate etc. as growth promoter and prebiotic have been studied earlier for having substitute to antibiotic.

Paul *et al.* (2012) revealed the effect of yeast cell wall preparation (of *Saccharomyces cerevisiae* origin) as an immunomodulator of the innate immune response. To evaluate its effect on chicken, yeast cell wall preparation (Nutriferm™) was administrated orally to 1 week old broilers @ 0.4gm and 0.8gm per kg feed for 15 days and then switched back to control diet for 20 days. Similar number of birds was kept separately as control with normal feeding regime. Non-specific immunity was assessed in randomly selected treated and control birds at the end of the experimental feeding period ('0' day) by performing neutrophil, monocyte and lymphocyte functional assay *in vitro*. In both dose group super oxide anion production by neutrophil was increased gradually up to '10' day post treatment (DPT) and then decreased to '20' DPT. The increase was significant (P<0.5) compared to control birds. In vitro nitrite production by monocyte was found to be high in treated birds than control. In 0.4gm treatment group in vitro non-specific lymphocyte proliferation and IL-2 production was first increased and then decreased abruptly. But in 0.8gm treatment group in vitro non-specific lymphocyte proliferation and IL-2 production was increased and then decreased gradually. The findings of the study showed that 15 days oral administration of yeast cell wall preparation on both the doses improve innate immune responses in the broiler chicks.

Xylanase is the name given to a class of enzymes which degrades the linear polysaccharide beta-1, 4 xylan to xylose, thus breaking down hemicelluloses which are a major component of the cell wall of the plants. Xylanases are known to increase protein digestibility of wheat and this is attributed particularly to release of protein from the xylan enriched aleurone layer. Xylanase supplementation improves conjugated bile acid function in intestinal contents and increase villus size of small intestine wall in broiler. Supplementing broiler diets with combination of xylanase and â-glucanase improves the nutritive value of the diet (Veldman and Vahl, 1994). The addition of xylanase improves weight gain, feed intake, feed efficiency, AME and decreased water intake (Wu *et al.*, 2004) and Vitamin E content of liver in broiler was significantly improved by addition of xylanase (Danicke *et al.*, 1999; 2001). Nutri-xylanase is a bacterial xylanase processed from *Bacillus subtilis* and produced by a micro-filtration advanced fermentation technique (Ganguly, 2013c).

Bar *et al.* (2012) conducted an experiment to study the effect of xylanase enzyme on body parameters of broiler birds. The experimental birds were distributed into five equal groups including control. Studies on body weight gain revealed that at the end of 4th week of experiment, significant difference in body weight gain among the birds of the control and various treatment group was noticed (P<0.05) though there was subtle difference among the birds of control and various treatment groups (T_1-T_4). Birds of T_3 group revealed the highest body weight gain followed by the birds of T_1, T_2 and T_4 groups respectively. Birds of the control group revealed the lowest body weight gain. At the end of 5th and 6th week of experiment there was significant difference (P<0.05) in the body weight gain among the birds of the control and various treatment group. Birds of T_3 group revealed the highest body weight gain and the lowest body weight gain were observed in the birds of control group.

Paul *et al.* (2013) studied the effect of purified α-glucan from an edible mushroom (*Pleuratus florida*) as an immunomodulator on the innate immune responses in broiler chicken (40 No.) , purified mushroom glucan was administered orally to 1 week old broiler chicks (20 No.) @ 20 mg/kg feed for 15 days and then switch back to control diet. Similar number of birds was kept separately as control with normal feeding regime. Non-specific immunity and protective ability were assessed in treated and control birds at the end of the experimental feeding period (0 day) by performing neutrophil, macrophage and lymphocyte functional assay *in vitro* and challenged with virulent field isolate of Newcastle Disease. Superoxide anion production by neutrophil, *in vitro* non-specific lymphoproliferation and IL-2 production were increased gradually up to 10 days post treatment and the increase was highly significant (P<0.05) compared to control birds. *In vitro* nitrite production by macrophage was found to be high in treated birds. Also mushroom glucan as a feed supplement significantly provided protection against Newcastle Disease. The result showed the potentiality of α-glucan (mushroom origin) as an immunostimulant in poultry.

Mode of Immunomodulation

Prebiotics have the potentiality to enhance many host biological responses and reduce the mortality of fishes caused by invasion by pathogens. However, the anaerobic intestinal tract microbiota of commercially important fishes, such as channel catfish, hybrid striped bass, tilapia and salmonids need to be investigated to determine if there are any particular bacterial species to be enhanced with the use of prebiotics. By increasing the production of volatile fatty acids (VFAs) in the gastrointestinal (GI) tract, hosts benefit by recovering some of the lost energy from indigestible dietary constituents and by inhibiting potential pathogenic bacteria (Manning and Gibson, 2004; Vazquez *et al.*, 2005). The VFAs produced are also indicative of the microbial population present in the GI tract (Nisbet, 2002).

An immunomodulator is a substance (*e.g.* a drug) which has an effect on the immune system. Different Immunomodulators stimulate the immune system. Biological activities of immunomodulator are influenced by different physicochemical parameters, such as solubility, primary structure, molecular weight, branching and polymer charge (Bohn and BeMiller, 1995). During the development of immune reactions, immunomodulating the effects of α-glucans have been well established (Vetvicka and Sima, 2004).

There are two main categories of immunostimulators: a) Specific immunostimulators provide antigenic specificity in immune response, such as vaccines or any antigen. For specific immune response, hosts should have prior exposure to an antigen after which recognition and subsequent activation occurs through a coordinated action of B-lymphocytes and T-cells. B cells are lymphocytes that play a large role in the humoral immune response (as opposed to the cell-mediated immune response, which is governed by T cells. b) Non-specific immunostimulators act irrespectively of antigenic specificity to enhance the immune response of other antigens or to stimulate components of the immune system without antigenic specificity, such as glucans, synthetic drug levamisole etc. Many endogenous substances are non-specific immunostimulators. For example, Glucans and mannans possess non-specific immunostimulatory effect. α-glucan is a polymer of glucose consisting of linear backbone of α-1,3 linked D-glucopyranosyl residues with varying degree of branching from the C_6 position (Bohn and Bemiller, 1995). α-glucans are major components of yeasts, mushrooms and fungal mycelia. Mannan is a plant polysaccharide that is a polymer of the sugar mannose. Detection of mannan leads to lysis in the mannan binding lectin pathway.

Immunostimulants as a feed additive significantly provides protection against pathogen and upregulates phagocytosis, bacterial killing and oxidative burst.

Probiotic microorganisms in the gut stimulate the immune response of host system in within two ways (De *et al.*, 2009). They can migrate through the gut wall as viable cells thereby multiplying to a limited extent or the released antigens by the dead organisms are absorbed and stimulate the immune response directly. Probiotics generally find their applications in poultry feed because of their positive effects growth rate, improved feed conversion and improved resistance to diseases. Probiotics have a positive effect on host immune response through increased activity of macrophages shown by enhanced ability to phagocytose organisms or carbon particles, increased production of systemic antibody *e.g.* IgM and interferon and increased effect of local antibody at mucosal surfaces such as gut wall. The effect of probiotics on the host immune system can be measured by estimating the levels of macrophage enzymes.

Prebiotics have many beneficial effects such as increased disease resistance and improved nutrient availability (De *et al.*, 2009). As such, prebiotics they have the potential to increase the efficiency and sustainability of livestock and poultry production.

The most commonly used prebiotic preparations are fructooligosaccharide (FOS), transgalactooligosaccharide (TOS), inulin, glucooligosaccharide, xylooligosaccharide, isomaltooligosaccharide, soybeanoligosaccharide, polydextrose, lactosucrose (Vulevic *et al.*, 2004; Propulla, 2008). Natural sources of prebiotics in vertebrates include chicory, onion, garlic, leek, tomato, honey etc.

Activity on Immunostimulation

Glucans having a strong immunomodulating activity have been well studied in fishes and livestock (Anderson, 1992). Only a few immunostimulants can be used for commercial farming purposes (Siwicki *et al.*, 1998). Many studies have been carried out to measure the effects of glucans on host immunity. Some investigators have adopted the in vitro culture of macrophages with glucan (Cook *et al.*, 2001), but in vivo studies have been carried out by the majority of workers (Sahoo and Mukherjee, 2001; Ortuno *et al.*, 2002). Under intensive conditions, individuals are more susceptible to microbial infections especially in their growing stages (Smith et al., 2003). During stress, immunostimulants can provide resistance to pathogens. There are two types of glucans á- and â- glucans the numbers of which clarify the type of O-glycosidic bond. Glucans are commercially significant as immunostimulating agents. Different types of â-glucans have been used successfully to increase resistance of poultry, fish and crustaceans against bacterial and viral infections (Paulsen *et al.*, 2001; Bagni *et al.*, 2005; Ganguly *et al.*, 2009; Ganguly and Mukhopadhayay, 2011; Paul *et al.*, 2013). It has been seen that health, growth and general performance of livestock and poultry may be improved by the use of â-glucans (Paul *et al.*, 2012). Product source, animal species, development stage of the target organism, dose and type of glucan, route, time schedule of administration and the association with other immunostimulants affect the immunomodulatory effects of glucans (Guselle *et al.*, 2007).

The immunostimulatory effects of glucan, chitin, lactoferrin, levamisole, vitamins B and C, growth hormone and prolactin have been reported. These immunostimulants mainly facilitate the function of phagocytic cells and increase their bactericidal activities. Several immunostimulants also stimulate the natural killer cells, complement, lysozyme and antibody responses of host. The most effective method of administration of immunostimulants is by injection. Oral and immersion methods have also been reported, but the efficacy of these methods decreases with long-term administration. Overdoses of several immunostimulants induce immunosuppression. Growth promoting activity has been noted in individuals treated with glucan or lactoferrin. Immunostimulants can overcome immune suppression by sex hormones (Ganguly *et al.*, 2009; 2010).

For the effective use of immunostimulants, the timing, dosages, method of administration and the physiological condition of the host needs consideration. Immunostimulants can reduce the losses caused by disease in commercial poultry rearing; however, they may not be effective against all diseases (Ganguly *et al.*, 2009; 2010).

Physiological Implications

Eidelsburger and Kirchgessner (1994) reported that calcium formate alone or in combination with other acids when given at the rate of 0.5% and 1.5%, increased FCR and growth performance in broiler chickens up to 35 days of age. Benedetto (2003) also observed mix of organic acids (ACIDLAC) used as a replacer of AGPs and improved production performance in breeding hens along with other beneficial effects. Mairoka *et al.* (2004) also reported that mixture of organic acids, as a substituted of AGPs improved the performance of birds even in absence of antibiotic. Savage *et al.* (1997) concluded from a dose responsive study (0-0.33%) that MOS @ 0.11%, maximized weight gain in poultry up to 0-8 weeks of age. Stanley *et al.* (2000) found same type of effect with supplementation of 0.1% MOS on hybrid Tom's body weight gain. Parks *et al.* (2001) reported from a study from turkeys supplemented with MOS that MOS may be utilized as a alternative to AGPs to improve turkey performance.

It has already been reported that 1% formic acid or 1.45% calcium formate did not affect live weight of broiler chicken (Izat *et al.*, 1990). It was found out that 80% formic acid and 20% propionic acid mixture added at 1% level to broiler chicken ration did not affect live weight (Kaniawati *et al.*, 1992). It has also been reported that formic acid and propionic acid mixture (85% and 15%) added at 1% level to the broiler chicken ration in the initial period did not affect weight gain (Visek, 1978). Reports have also been made about significant increased in body weight gain with the supplementation of 0.5% lactic acid in drinking water (Veeramani *et al.*, 2003). It was also revealed that increased in body weight with supplementation of lactic acid. The mix of organic acids improves performance of birds (Maiorka *et al.*, 2004). From a dose responsive study it was concluded (0-0.33%) that MOS @ 0.11%, maximized weight gain in poultry up to 0-8 weeks of age. The same type of effect was found with supplementation of 0.1% MOS on hybrid Tom's body weight gain (Parks *et al.*, 2001) conducted in turkeys supplemented with MOS that MOS may be utilized as a alternative to AGPs to improve turkey performance (Podolsky, 1995; Pelicano, 2003).

Body Performance

Dietary supplementation of different feed additives e.g. immunostimulants, probiotics and prebiotics usually in small quantities for the purpose of fortifying it with certain nutrients have been found to be beneficial for improving immune status, feed efficiency and growth performance of living beings also. Ganguly *et al.* (2012) highlighted the effect of dietary prebiotics on different body growth parameters of fish. However, the application of prebiotics and supplementary enzymes in fish feed is now gradually gaining importance in commercial livestock farming practices, the article stresses on the effect of prebiotics on live body weight gain, dressing percentage, weight of vital organs and muscles and mean villus lengths in digestive tract of fish along with their application as growth promoters in commercial poultry and aquafeed (Gatesoupe, 2005, Ganguly *et al.*, 2010).

Mannion (1981) reported that the body weight gain was improved by 12 to 25% and feed consumption was increased by 3 to 21% when chicks at 4 weeks of age fed diets supplemented with enzyme like xylanase. Veldman and Vahl (1994) noticed xylanase supplementation improved food conversion ratio by 2.2-2.9% and body weight gain by 0.2-2.5%. Silverslides and Bedford (1999) showed xylanase supplementation had a positive body weight gain and the feed to gain ration. Danicke *et al*. (2001) found addition of xylanase significantly increased the weight gain up to 21 days of age and decreased the feed to gain ratio slightly. Mathlouthi *et al*. (2003) showed feed efficacy and body weight gain was improved with the supplementation of xylanase. Wu *et al*. (2004) observed that the xylanase supplementation significantly improved weight gain, feed efficiency and AME. Wu *et al*. (2004) observed that addition of xylanase and phytase reduced the relative weight gain of the small intestine by 15.5% and 11.4% respectively. Yubo *et al*. (2005) reported body weight and feed per gain ratio was improved (P<0.05) by xylanase supplementation in the first 2-3 weeks in broilers. Ahmad *et al*. (2007) noticed xylanase supplementation increased body weight, feed intake and feed gain ratio. Liu *et al* (2007) observed xylanase supplementation increased body weight gain from 0-21days of age of broilers. Gao *et al*. (2008) found that the supplementation of xylanase enzyme improved (P<0.05) growth performance and feed conversion efficiency (FCE).

Pelicano (2005) reported that there was effect on the dressing percentage and weight of different organs and muscles at 21 and 42 days without any major influence on the dressing percentage, organ and muscle weight under different treatment groups with organic acid salts individually and its combination in broiler birds. Higher villus height in duodenum, jejunum in small intestine was reported with most organic acidifier in diet of broiler (Loddi *et al*., 2004). Again it was reported higher villus height in the ileum with the diet based on organic acidifier compared with diet fed without MOS + organic acidifier (Savage *et al*., 1997). Therefore, the supplementation of organic acidifier may increase villus height of different parts of small intestine. So organic acidifier reduces the growth of many pathogenic and non-pathogenic intestinal bacteria, decreases intestinal colonization and reduces infections process, ultimately decreasing inflammatory process at the intestinal mucosa. It increases villus height and function of secretion, digestion and absorption of nutrients can be appropriately performed by the mucosa (Iji and Tivey, 1998). It was also reported positive effects of the use of prebiotics on the intestinal mucosa among which a significant increase in villus height of three segments of small intestine of birds, age one week and supplemented with MOS (Maiorka *et al*., 2004). In the present study, use of organic acidifier in diet significantly increased villus height of different segment of small intestine of broiler possibly by reducing intestinal colonization of pathogenic and non-pathogenic bacteria, which is complied in the findings of Savage *et al*. (1997) and Loddi *et al*. (2004) respectively.

Loddi (2003) described higher villi in the intestinal mucosa (duodenum) of birds fed with MOS at 7 and 21 days of age respectively. Pelicano *et al*. (2005) reported that in jejunum MOS + OA resulted in higher villi in the jejunum (P<0.01) followed by the diets containing MOS based prebiotics While In case of ileum present investigation was partially corroborated with the findings of Pelicano *et al*. (2005) who observed higher villi length when birds were fed with prebiotic based on MOS, compared to the control group.

Microorganisms that is sensitive to acid pH and results in higher villi length (Radecki and Yokoyama, 1991). Some bacteria may recognize binding sites on the prebiotics instead of intestinal mucosa and the colonization by pathogenic bacteria in intestine is thus reduced. Therefore, besides a lower infection incidence, there is an increase in the absorption of available nutrients, a mechanism that directly affects the recovery of the intestinal mucosa, increasing villi length. These results disagree to those obtained by Pelicano *et al.* (2003) and Santin *et al.* (2001) respectively, who found no difference in ileal villi length with the use of probiotics and prebiotics.

Roy *et al.* (2012) conducted experimental studies to evaluate the pathomorphological effect of different combination of organic acids *viz.* formic acid, propionic acid and lactic acid as a replacer of growth promoter antibiotics. The birds were divided into six equal groups *viz.* negative control (C_1), positive control (C_2) and treatment groups (T_1, T_2, T_3 and T_4). Birds of group C_1 were supplied with diet without any antibiotics or acid, group C_2 with Virginiamycin @ 500 gm/quintal feed, group T_1 with 0.3% ammonium formate group T_2 with 0.3% calcium propionate, group T_3 with 0.15% ammonium formate and 0.15% calcium propionate and group T_4 with 0.1% ammonium formate, 0.1% calcium propionate and 0.1% calcium lactate. Body weight gain was higher in C_1 and C_2 compared with treatment groups in first two weeks, but pathological changes were maximum in negative control, *i.e.* after 6 weeks weight gain was significantly better in T_3 and T_4 than in groups C_1, T_1 and T_2 groups. Pathomorphological changes in group C_1 and T_2. Group C_2 and T_1 also showed same types of changes and in group T_3 and T_4 the changes were less. The most common changes among the groups were thickening of muscular layer, accumulation of inflammatory cells and congested blood vessels. In T_1 group, rupture of villi was also seen. In T_3 and T_4 the intestinal glands were hypertrophied. Based on present investigation, it is suggested that combination of organic acid may be used in broiler feed as a replacer of growth promoter antibiotic instead of using only one acid.

It was reported that combined use of MOS and organic acid salts in poultry feed can be used as an alternative to the antimicrobial and antibiotic growth promoters to achieve good health for sustainable and economic poultry production in India (Das *et al.*, 2012).

Microbial Biomass Load in Gastrointestinal Tract

The findings that MOS and OA successfully reduces bacterial load in the intestine of broiler birds were in accordance with the findings of Newman (1994), Lon (1995), Spring *et al.* (2000) and Fairchild *et al.* (2001). Stanley *et al.* (2004) concluded that yeast cell culture residue (YCR) treated broiler chicks resulted lower intestinal Coliform population in comparison to control and other antibiotic treated (lasalocid @ 90.7kg/ton, bacitracin @50gm/ton) groups. Sims *et al.* (2004) also found that MOS + BMD treated turkeys showed significantly lower *Clostridium perfringens* population in the gut than control at 6 weeks of age. MOS is believed to block type-I fimbriae and prevents pathogens from attaching to the intestinal lining and passes them out of the gut. (Dawson and Pirulescu, 1999). Bacteriological studies of intestine of broiler birds revealed reduced *E. coli* compared with the untreated control group. Salmonellae were not found in any group. No significant difference was observed among control and treatment groups regarding the number of *Clostridium* sp. present in the intestine. The above result clearly indicated that consumption of prebiotics mixed with poultry feed reduced the load of coliform bacteria in the intestine (Roy *et al.*, 2012).

The pharmaceutical and physiological effect of purified α-glucan from an edible mushroom (*Pleuratus florida*) as an immunomodulator is immense on the innate immune responses in broiler Also, mushroom glucan as a feed supplement significantly provides protection against disease. This article portrays the potentiality of α-glucan (mushroom origin) as an immunostimulant.

Immunomodulator stimulates leucocytes, particularly cells of the macrophage system and modulates and potentiates the immune system of the body (Seljelid, 1990). It has been recommended earlier that the constant addition of immunomodulators to feed is beneficial for prevention of diseases (Onarheim, 1992). One of such immunostimulant compound is α-Glucan, polymers of glucose which consists of a linear backbone of α-1, 3 linked D- glucopyranosyl residues having varying degree of branching from the C_6 position (Bohn and BeMiller, 1995). α- Glucans are major structural components of yeast, mushrooms and fungal mycelia. Supplementation of α- glucan in diets increase the macrophage phagocytic activity, PHA-P- mediated lymphoproliferative response and also humoral response (Guo et al., 2003). α- Glucan provides significant protection against pathogen as a feed additive by upregulating phagocytosis, bacterial killing, and oxidative burst in chicken (Lowry *et al.*, 2005). In the mammalian system, action of α- glucan is mediated through toll-like receptors (TLR) and dectin-1 (Lowry *et al.*, 2005). In the present work evaluation was carried out for short term dietary influence of a purified α- glucan, prepared from an edible mushroom, on the innate immunity and disease resistance of broiler birds.

Immunomodulator is a substance that stimulates leucocytes-particularly cells of the monocyte/ macrophage system and thereby modulates, and most often potentiates, the immune system of the body (Seljelid, 1990). The term immunomodulator was often used interchangeable with immunostimulants, adjuvants and biological response modifiers. Glucan and mannan are the main components of yeast cell wall (YCW) that are gained from pure culture of yeast, *Saccharomyces cevisiae*. α-D- glucan is major component of yeast cell wall and has been shown to stimulate non-specific immune response. Glucans with α-1,3, α-1,4 and α 1-6 glucosidic linkages are major structural components of YCW (Brown and Gordon, 2003), mice (Selvaraj *et al.*, 2005), rats (Williams and, Diluzio, 1979), rabbits Reynolds *et al.*, 1980), sheep and pigs (Xiao *et al.*, 2004).

Importance as Dietary Supplement

Yeast α-glucan has been reported to enhance the immune responses in fish (Ganguly, 2009; 2010; 2013a;b), cattle (Persson Waller *et al.*, 2003) and humans (Engstad *et al.*, 2002). However, information regarding the effect of dietary administration of yeast cell wall preparation on immune responses in birds is limited. In the present study we evaluate the augmentation of the non-specific immune responses, *viz.*, production of oxygen and nitrogen species, lymphoproliferation and IL-2 (cytokine) production in broiler birds following YCW treatment.

Previous studies showed that infections caused by *Staphylococcus aureus* and *Eimeria vermiformis* in mice can be prevented by α-glucan administration (Yun *et al.*, 2003). Experimental respiratory challenge with *Escherichia coli* in broiler chicks can also be prevented by α-1,3 / 1,6 glucan derived from *Saccharomyces cerevisiae* (Huff *et al.*,

2006). Rice *et al.* (2005) showed that dietary administration of glucan to rat enhanced survivability against *Staphylococcus aureus* infections. Orally administered yeast â-glucan to mice could reduce the mortality in anthrax infections (Vetvika et al., 2002).

Immunomodulatory Implications

In vertebrates, the immunomodulating abilities of α-glucans are thought to stem from their ability to activate leukocytes, but there is some confusion about their precise biological effects (Brown and Gordon, 2003). Paul *et al.* (2012) assessed the immunostimulatory role of glucan extracted from yeast (*Saccharomyces cerevisiae*) cell wall was assessed in two different doses in terms of cellular immune effector activity. The production of oxygen radicals by YCW (both dose group) fed broiler birds was higher up to 20th day post treatment than control values. The O.D. value was in peak level at 10th day post treatment and significantly higher than control group (P<0.05) and then the O.D. values on 20th day decreased. The oxygen radical production in 0.8gm/kg treatment group was higher than 0.4 gm treatment group on 10th day post treatment. Nitrite production was increased in both YCW fed groups than control group at 0 day (Paul *et al.*, 2013). From 10th day onward the nitrite production level was decreased in 0.8 gm treatment group but in 0.4gm treatment group nitrite production was peak level at 10th day post treatment. In 0.4gm treatment group in vitro non-specific lymphocyte proliferation and IL-2 production was first increased and then decreased abruptly. But in 0.8 gm treatment group *in vitro* non-specific lymphocyte proliferation and IL-2 production was increased and then decreased gradually and IL-2 production was in peak level at 10th day post treatment (Paul *et al.*, 2013).

The previous workers showed that the use of yeast glucan was enhanced oxidative respiratory burst in human and chicken (Wakshull *et al.*, 1999), monocyte activity and nitrite production also enhanced in sheep and chicken (Waller and Colditz, 1999).

Guo *et al.* (2003) and Waller *et al.* (1999) observed glucan enhanced the lymphocyte proliferation in cattle. Oral administration of yeast glucan enhanced the cytokine production in mice (Tsukada *et al.*, 2003). The enhancement of oxygen radicals, nitrite, cytokine (IL-2) production and lymphoproliferation of broiler birds might be related to the oral administration of yeast cell wall preparation (Nutriferm™) from *Saccharomyces cerevisiae*.

The pharmaceutical and physiological effect of purified α-glucan from an edible mushroom (*Pleuratus florida*) has been extensively explored as an immunomodulator on the innate immune responses in broiler Also, mushroom glucan as a feed supplement significantly provides protection against disease. This article portrays the potentiality of α-glucan (mushroom origin) as an immunostimulant in poultry. Plant derived and herbal feed additives (often also called phytobiotics or botanicals) are commonly defined as plant-derived compounds incorporated into diets to improve the productivity of livestock through amelioration of feed properties, promotion of the individual production performance, and improving the quality of food derived from those animals, such as herbs (flowering, non woody, and non persistent plants), spices (herbs with an intensive smell or taste commonly added to human food), essential oils (volatile lipophilic compounds derived by cold expression or by steam or alcohol distillation), or oleoresins (extracts derived by non aqueous solvents). Cow urine therapy and all traditional practices from Indian systems of medicine have a strong scientific base. Traditional systems in medicines, whether from Ayurveda or Siddha or the use of cow urine distillate as immunomodulator are based on classical texts and systems, practices and products handed down over generations going back to Charak, Sushrutha, Vagabhatta, the Ashtangahridaya and the Samhitas. Cow urine has been described in 'Sushrita Samhita' and 'Ashtanga Sangraha' to be the most effective substance/secretion of animal origin with innumerable therapeutic values. In Ayurveda cow urine is suggested for improving general health. The present article highlights and portrays the immunopotential effect of CUD and CUD can be recommended in broiler ration at optimum dose level against NDV.

Immunomodulator stimulates leucocytes, particularly cells of the macrophage system and modulates and potentiates the immune system of the body[1]. It has been recommended earlier that the constant addition of immunomodulators to feed is beneficial for prevention of diseases. One of such immunostimulant compound is â-Glucan, polymers of glucose which consists of a linear backbone of α-1, 3 linked D-glucopyranosyl residues having varying degree of branching from the C_6 position[3]. â-Glucans are major structural components of yeast, mushrooms and fungal mycelia. Supplementation of α- glucan in diets increase the macrophage phagocytic activity, PHA-P- mediated lymphoproliferative response and also humoral response. α- Glucan provides significant protection against pathogen as a feed additive by upregulating phagocytosis, bacterial killing, and oxidative burst in chicken[5]. In the mammalian system, action of α- glucan is mediated through toll-like receptors (TLR) and dectin-1. In the present work evaluation was carried out for short term dietary influence of a purified α- glucan, prepared from an edible mushroom, on the innate immunity and disease resistance of broiler birds.

Immunomodulator is a substance that stimulates leucocytes-particularly cells of the monocyte/ macrophage system and thereby modulates, and most often potentiates, the immune system of the body. The term immunomodulator was often used interchangeable with immunostimulants, adjuvants and biological response modifiers. Glucan and mannan are the main components of yeast cell wall (YCW) that are gained from pure culture of yeast, *Saccharomyces cevisiae*. â-D- glucan is major component of yeast cell wall and has been shown to stimulate non-specific immune response. Glucans with α 1-3, α 1-4 and α 1-6 glucosidic linkages are major structural components of YCW, mice, rats, rabbits, sheep and pigs.

The phytogenic growth promoters supplemented in the diet or added in the drinking water in the broiler birds have a promising biological effect on their growth performance, to reduce the pathogenic bacteriological load in different parts of digestive tract and to increase villus height in different segments of small intestine mainly in duodenum.

Within phytogenic feed additives, the content of active substances in products may vary widely, depending on the plant part used (*e.g.* seeds, leaf, root or bark), harvesting season, and geographical origin. The technique for processing (*e.g.* cold expression, steam distillation, extraction with non aqueous solvents etc.) modifies the active substances and associated compounds within the final product.

Experimentally, it has also been proved that among urine from various species the urine of the Indian cows is most effective[11] for its medicinal properties. Immunomodulation is gaining importance for immunopotentiation in hosts against various infections. The cow urine distillate (CUD) is found to have immunomodulatory effect in mice as it enhances both T-and B-cell proliferation and also increases the level of IgG. Recently, the cow urine has also been granted U.S. patents (No. 6896907 & 6410059) for its synergistic properties with antibiotics, antifungal and anti-cancer drugs as bio-enhancer. It has provided the base for further research on immunomodulatory properties of indigenous cow urine. It has also been reported that CUD enhances B and T lymphocyte blastogenesis, increases IgG antibody titer in avian species. Keeping in view all the above facts, the present investigation was planned to study the immunomodulatory effect of cow urine distillate on humoral and cell mediated immune response against NDV vaccination in broiler chicks when administered orally.

Importance as Dietary Supplement

Yeast α-glucan has been reported to enhance the immune responses in fish, cattle and humans. However, information regarding the effect of dietary administration of yeast cell wall preparation on immune responses in birds is limited. In the present study we evaluate the augmentation of the non-specific immune responses, *viz.*, production of oxygen and nitrogen species, lymphoproliferation and IL-2 (cytokine) production in broiler birds following YCW treatment.

Studies showed that infections caused by *Staphylococcus aureus* and *Eimeria vermiformis* in mice can be prevented by α-glucan administration. Experimental respiratory challenge with *Escherichia coli* in broiler chicks can also be prevented by α-1,3 / 1,6 glucan derived from *Saccharomyces cerevisiae*. Rice *et al.* showed that dietary administration of glucan to rat enhanced survivability against *Staphylococcus aureus* infections. Orally administered yeast α-glucan to mice could reduce the mortality in anthrax infections.

The phytogenic growth promoter remains active throughout the gastrointestinal tract and as a consequence, it will exert broad spectrum antimicrobial action, will enhance nutrient utilization by exhibiting improvement in overall growth performance of broilers and by augmenting the gastrointestinal histomorphology thereby enhancing the host immunity.

Immunomodulatory effect of cow urine or its distillate has been reported by many workers and therefore this has made the base for present research. The dose of CUD selected in the present study is according to the recommendation by Kumar *et al.*

Jojo *et al.*[31] documented that the levamisole treated group of chicks also showed significant effect on MHI antibody titer in comparison to CUD suggesting its superior immunopotentiating effect over CUD on humoral immune response upon vaccination.

Awadhiya *et al.*, Srikumar *et al.*, Kumari and Rakhi showed increased cell mediated immune (CMI) response correlated with the findings. The findings were also in accordance with those of Chauhan et al., Ambwani and Garg et al. who worked on lymphocytes blastogenic activity with respective mitogens using lymphocyte proliferation assay.

Implications in Immunomodulation and Body Growth Promotion with Influence on Hematological and Biochemical Parametrs

In vertebrates, the immunomodulating abilities of α-glucans are thought to stem from their ability to activate leukocytes, but there is some confusion about their precise biological effects. Paul *et al.* assessed the immunostimulatory role of glucan extracted from yeast (*Saccharomyces cevisiae*) cell wall was assessed in two different doses in terms of cellular immune effector activity. The production of oxygen radicals by YCW (both dose group) fed broiler birds was higher up to 20th day post treatment than control values. The O.D. value was in peak level at 10th day post treatment and significantly higher than control group (P<0.05) and then the O.D. values on 20th day decreased. The oxygen radical production in 0.8gm/kg treatment group was higher than 0.4 gm treatment group on 10th day post treatment. Nitrite production was increased in both YCW fed groups than control group at 0 day[39]. From 10th day onward the nitrite production level was decreased in 0.8 gm treatment group but in 0.4gm treatment group nitrite production was peak level at 10th day post treatment. In 0.4gm treatment group in vitro non-specific lymphocyte proliferation and IL-2 production was first increased and then decreased abruptly. But in 0.8 gm treatment group in vitro non-specific lymphocyte proliferation and IL-2 production was increased and then decreased gradually and IL-2 production was in peak level at 10th day post treatment.

The previous workers showed that the use of yeast glucan was enhanced oxidative respiratory burst in human and chicken, monocyte activity and nitrite production also enhanced in sheep and chicken.

Guo *et al.* and Waller *et al.* observed glucan enhanced the lymphocyte proliferation in cattle. Oral administration of yeast glucan enhanced the cytokine production in mice[42]. The enhancement of oxygen radicals, nitrite, cytokine (IL-2) production and lymphoproliferation of broiler birds might be related to the oral administration of yeast cell wall preparation (Nutriferm™) from *Saccharomyces cevisiae*.

Burt stated microbial analysis of minimum inhibitory concentration (MIC) of plant extracts from spices and herbs, as well as of pure active substances revealed levels that considerably exceeded the dietary doses when used as phytogenic feed additive. Aksit *et al.* reported antimicrobial action of phytogenic feed additive may be in improving the microbial hygiene of carcass.

Batal and Parsons indicated that micronutrients also influenced the morphology of intestines. They observed an increased height of villi of jejunum in broilers at 28th day of age when fed with 5 g BioMos/kg from 7 to 28 day. Jamroz *et al.* have conducted a study that phytogenic formulations contained pungent principles (*e.g.* capsaicin) significantly increased intestinal mucus production.

Jamroz and Kamel observed on the improvements in daily weight gain (8.1%) and in feed conversion ratio (7.7%) of chickens when feed with diets supplemented (300mg/kg) with a plant extract containing capsaicin, cinnamaldehyde and carvacrol. Biavatti *et al*. reported *Alternanthera brasiliana* extracts (180 ml/200 kg feed) improved broiler performance from 14 to 21 days. Hernandez *et al*. studied that blend of essential oils of cinnamon, pepper and oregano compounds improved digestibility of nutrients in chicken. Jang *et al*. in chicken is the benefit of some natural substances on gastro intestinal enzymatic activity, most likely improving nutrient digestibility.

An experiment was conducted for evaluating the efficiency or effect of the phytogenic growth promoter. The phytogenic growth promoter was active throughout the gastrointestinal tract and as a consequence, it will exert broad spectrum antimicrobial action, will enhance nutrient utilization by improving gastrointestinal absorptive properties and will augment the host immunity. In the experiment, two proven and approved phytogenic growth promoters, Digestarom 1317 (dosage 150 ppm) and Digestarom 1440 (dosage 800 ppm) AC were fed to the broiler chickens against an antibiotic growth promoter, Bacitracin Methylene Disalicylate (BMD).

Digestarom AC is a combination of phytogenic components with glycerides of short chain fatty acids. Basically, Digestarom AC is a complex of plant extracts and plant essential oils along with monoglycerides, lactic acids and multiglyceride complexes. Being a complex of plant extracts and essential oils, Digestarom AC is hypothesized to stimulate feed intake, intestinal secretion of enzymes and enhance digestibility of nutrients. Additionally, Digestarom AC is anticipated to act as a broad spectrum antimicrobial substances throughout the gastrointestinal tract and promote development of the villus structure of the gut.

Combining cow urine distillate (the term 'distillate' itself is a misnomer, since the material used is the residue, not the distillate) with antibiotics is not recommended at all and its combination in liquid or lyophilized powder form with modern drugs is irrational, since the relative bioavailability and pharmacokinetics of the components remain unknown. In vitro experiments with cow urine distillate have little relevance, since activity in vivo largely depends on plasma levels, which in turn are related to serum binding properties and absorption. Mammalian urine contains useful constituents like adrenocorticotropic hormaone (ACTH) isolated from pregnant female urine. Other constituents include various enzymes, amino acids and Erythropoetin. The reported results of experiments which have been carried out on cow urine distillate in India and the grant of the U.S. patent vindicates the use of cow urine as a bio-enhancer. According to a recent online report of 'Love4Cow Trust', researchers at Central Institute of Medicinal and Aromatic Plants (CIMAP), Lucknow have identified a fraction of cow urine distillate as bio-enhancer of commonly used antibiotics and anti-cancer drugs. Bio-enhancers do not possess drug activity of their own but promote and augment the bioactivity or bioavailability or the uptake of drugs in combination therapy. Such bio-enhancers have been earlier isolated only from plant sources. In the study at CIMAP, Lucknow, researchers found that 'cow urine distillate fraction' enhances the activity of antibiotics such as rifampicin by about 5-7 folds against *E. coli* and 3-11 folds against Gram-positive bacteria. Rifampicin is a front-line anti-tubercular drug used against tuberculosis. Interestingly, it was also found that 'cow urine distillate fraction' enhanced the potency of 'Taxol' (paclitaxel) against MCF-7 a human breast cancer cell line in *in-vitro* assays (US Patent No.6,410,059).

The dietary â-glucan may provide immunostimulatory properties necessary to reduce the incidence of any infection in poultry. Cow urine distillate (CUD) possesses immunomodulatory effect as judged by increase in HI antibody titer against viral infection. The immunopotentiating effect of CUD has been analysed on humoral and cell mediated immune response with virulent virus vaccination, its use as an immunomodulating agent at proper dose level may be advocated. The phytogenic growth promoter enhance productive performance of the broiler in terms of body weight gain with minimum alteration of gut morphology and the possibility of bacterial invasion is much less. Phytogenic growth promoter can be used as a potent replacer of antibiotic growth promoter if used at optimum level.

Phytogenic feed additives (often also called phytobiotics or botanicals) are commonly defined as plant-derived compounds incorporated into diets to improve the productivity of livestock through amelioration of feed properties, promotion of the individual production performance, and improving the quality of food derived from those animals, such as herbs (flowering, non woody, and non persistent plants), spices (herbs with an intensive smell or taste commonly added to human food), essential oils (volatile lipophilic compounds derived by cold expression or by steam or alcohol distillation), or oleoresins (extracts derived by non aqueous solvents).

The phytogenic growth promoters supplemented in the diet or added in the drinking water in the broiler birds have a promising biological effect on their growth performance, to reduce the pathogenic bacteriological load in different parts of digestive tract and to increase villus height in different segments of small intestine mainly in duodenum.

Within phytogenic feed additives, the content of active substances in products may vary widely, depending on the plant part used (*e.g.* seeds, leaf, root or bark), harvesting season, and geographical origin. The technique for processing (*e.g.* cold expression, steam distillation, extraction with non aqueous solvents etc.) modifies the active substances and associated compounds within the final product.

Influence and Impact on Physiological Parameters

The phytogenic growth promoter remains active throughout the gastrointestinal tract and as a consequence, it will exert broad spectrum antimicrobial action, will enhance nutrient utilization by exhibiting improvement in overall growth performance of broilers and by augmenting the gastrointestinal histomorphology thereby enhancing the host immunity.

Recent Investigations on the Related Aspect

Burt[1] stated microbial analysis of minimum inhibitory concentration (MIC) of plant extracts from spices and herbs, as well as of pure active substances revealed levels that considerably exceeded the dietary doses when used as phytogenic feed additive. Aksit *et al.* reported antimicrobial action of phytogenic feed additive may be in improving the microbial hygiene of carcass.

Batal and Parsons indicated that micronutrients also influenced the morphology of intestines. They observed an increased height of villi of jejunum in broilers at 28th day of age when fed with 5 g BioMos/kg from 7 to 28 day. Jamroz *et al*[l] have conducted a study that phytogenic formulations contained pungent principles (*e.g.* capsaicin) significantly increased intestinal mucus production.

Jamroz and Kamel observed on the improvements in daily weight gain (8.1%) and in feed conversion ratio (7.7%) of chickens when feed with diets supplemented (300 mg/kg) with a plant extract containing capsaicin, cinnamaldehyde and carvacrol. Biavatti *et al*. reported *Alternanthera brasiliana* extracts (180 ml/200 kg feed) improved broiler performance from 14 to 21 days. Hernandez *et al*. studied that blend of essential oils of cinnamon, pepper and oregano compounds improved digestibility of nutrients in chicken. Jang *et al*. in chicken is the benefit of some natural substances on gastro intestinal enzymatic activity, most likely improving nutrient digestibility.

The phytogenic growth promoter enhance productive performance of the broiler in terms of body weight gain with minimum alteration of gut morphology and the possibility of bacterial invasion is much less. Phytogenic growth promoter can be used as a potent replacer of antibiotic growth promoter if used at optimum level.

Honey is the organic, natural sugar, produced from the nectar and exudation of plant by honey bees, *Apis mellifera*. It is used as traditional food since the date back to 2100 B.C. Honey bees transform nectar of flowers into honey by the process of regurgitation and evaporation. The sweetness of honey comes from the monosaccharide (fructose and glucose). The carbohydrates of honey provide strength and energy to our body. Honey has important effect in instantly boosting endurance and it reduces muscle fatigue of the body. The glucose in honey gives immediate energy to body boost, where as the fructose is absorbed slowly by the body providing sustained energy. Honey has powerful immune system booster. Its antioxidant and antibacterial properties help to improve digestive system. Other than carbohydrates, honey contains proteins, enzymes, amino acids, minerals, trace elements, vitamin and other photo-chemical. Different types of honey have different compositions of biochemical and biophysical properties.

Different types of antioxidants such as phenolic acids and flavonoids, certain enzymes (glucose oxidase and catalase) are present in honey. Now these days the evidence of the antioxidant capacity of honey has been proved that honey can prevent deteriorative oxidation reaction in foods, such as lipid oxidation in meat and enzymatic browning of fruits and vegetables. So, honey is beneficial to serve as a natural food antioxidant. In the previous studies it was examined that honey is similar in antioxidant capacity to many fruits and vegetables on a fresh weight basis, which is measured by the Oxygen Radical Absorbance Capacity (ORAC) assay.

However, the antioxidant properties of honey, varies depending upon the floral sources. There is a lack of knowledge about these antioxidant properties of honey. Several studies on Indian honey have been shown that honey has a high phenolic profile consisting of benzoic acids and their esters, and flavonoid aglycans.

Honey is a traditional treatment for infected wounds as long as 2000 years ago when bacteria were discovered and also it can be effective on antibiotic resistant strains of bacteria. The antibacterial properties vary depending upon the region from where it has generally collected. Honey is selected for clinical testing which should be evaluated on the basis of antibacterial activity levels determined by laboratory testing. Hydrogen peroxide, which is known as antimicrobial agent, is produced in honey solution commonly used as antiseptic. The harmful effects of hydrogen peroxide are further reduced because honey inactivates the free ions which catalyses the formation of oxygen free radicals. Now these days, scientists are reporting that the antibacterial properties of honey are very useful for the treatment of wound healing. Different types of wound infections including burns, venous leg ulcers, leg ulcers of mixed etiology, diabetic foot ulcers, pressure ulcers, unhealed graft donor sites, abscesses, boils, pilonidal sinuses, infected wounds from lower limb surgery, necrotizing fasciitis and neonatal postoperative wound infections are treated by honey.

Honey an Ancient Medicine

Honey, a well known natural healing agent with antibacterial and antioxidant property was used by ancient Greek people for thousands of years to treat wound. Honey is a traditional treatment for infected wounds as long as 2000 years ago when bacteria were discovered and also it can be effective on antibiotic resistant strains of bacteria. It is used as traditional food since the date back to 2100 B.C. Honey is the organic, natural sugar, produced from the nectar and exudation of plant by honey bees, *Apis mellifera*. Honey bees transform nectar of flowers into honey by the process of regurgitation and evaporation.

Physical and Chemical Composition of Honey

Honey is basically acidic in nature. The pH and acidity level changes depending upon the botanical origin mainly geographic origin of honeys the level of pH stands in between 2-6. Honey contains minerals and acids serving as electrolytes, which can conduct the electrical current. The measurement of electrical conductivity (EC) was introduced in 1964. The average conductivity of Nigerian honey is between 9.419-172.900 is cm^{-1}.

Moisture content of honey depends upon harvest season, along with the degree of maturity reached in the live. Honey has water content between 15-18 %. Color in liquid honey varies from clear and colorless (like water) to dark amber or black. The color of honey is amber or gold. The range of color of honey is in between 52.00-255.00.

The average range of ash content of honey varies between 0.095-0.518.[1] The ash content of locally produced honey samples ranged between 0.047-0.35 which is within average standard limits. Ash content of honey is about 0.21 % of its weight, but it varies from 0.02-1.0 % (quality and standards authority of Ethiopia, 2005).

The natural product honey has been reported to contain about 200 substances, which consist of not only highly concentrated solution of sugars, but also the complex mixture of other saccharides, amino acids, peptides, enzymes, proteins, organic acids, polyphenols, carotenoid like substances, vitamins, and minerals. Sugars are the main constituents of honey, comprising about 95 % of its dry weight. While glucose and fructose are dominant constituents, among 25 different sugars have been detected. According to White (1975), protein present in honey is mainly enzymes. Honey contains roughly 0.5 % proteins and the protein contents in some honeys can be over 1000 μg/g). Main enzymes are diastase, invertase, glucose oxidase and catalase. Although the content of amino acids in honey is relatively small, it has been found that almost all of physiologically essential amino acids are present in honey.

The primary amino acid is proline, contributing 50-85 % of the total amino acids. The level of organic acids in honey is relatively low and about 18 organic acids have been detected. Most of the acidity present in honey is added by honeybees. Gluconic acids, the predominant organic acid, are the product of glucose oxidation, presenting at 50-fold higher levels than other acids. Investigations have shown that a wide range of trace elements are present in honey, including Al, Ba, Bi, Co, Cr, Mo, Ni, Pb, Sn, Ti as well as minerals (Ca, Cu, Fe, K, Na, Mg, Zn), among them, the main mineral element is potassium while copper present lowest amount. Vitamins such as thiamin (B_1), riboflavin (B_2), pyridoxine (B_6), and ascorbic acid (C) have also been reported but their amount is very small in honey. When honey is treated with mild heat or prolonged storage, a compositional change can occur due to caramelization of the carbohydrates, the Maillard reaction, and decomposition of fructose in the acid medium of honey.

Antioxidant Properties of Honey

Honey contains a significantly high level of antioxidants, both enzymatic and nonenzymatic, including catalase, phenolic acids, flavonoids, carotenoids, organic acids, ascorbic acids, amino acids and Maillard reaction products. Phenolic compounds commonly found in honey include phenolic acids, flavonoids and polyphenols. Honey phenolic acids can be proteocatequic acid, phydroxibenzonic acid, caffeic acid, chlorogenic acid, vanillic acid, p-coumaric acid, benzoic acid, ellagic acid, cinnamic acid, and flavonoids in honey consist of naringenin, kaempferol, apigenin, pinocembrin, chrysin, galangin, luteolin etc. The large and complex flavonoids greatly contribute to honey color, flavor, anti-fungal, and antibacterial activity. The antioxidant capacity of different honeys depends on the floral sources used by bees to collect nectar, seasonal and environmental factors, as well as processing ways. Although the total antioxidant activity of honey is the combination of a wide range of active substances, the content of phenolic compounds can significantly reflect the total antioxidant activity of honey to some extent. However, the level of phenolic compounds present in honey is not always positively proportional to its antioxidant activity. The explanation for this activity may be due to the presence of variable types of polyphenols, thereby providing variable scavenging activity. Darker honey is likely to have a higher antioxidant contents than light colored honey. As well, the antioxidant content is higher in honey with higher water content.

Use for Wound Healing

A review of honey's use in wound care by Molan (2006) has overwhelming evidence that honey is a credible wound treatment option. With regards to wound treatment by honey application, the osmotic action of honey can induce outflow of lymph, which is able to promote extra oxygenation and provide improved supply of nutrients on the wound surface, as well as to flush away proteases that may inhibit the repair process. Moreover, honey's osmotic action can create a moist environment that is required for the fibroblasts to contract and pull the margins of wound together. The acidic pH of honey also adds the value to aid wound healing since it can facilitate to release the oxygen carried by hemoglobin. It has been noted that acidification of wounds can improve the speed of the healing process. A number of studies have firmly reinforced that honey is an effective medicinal treatment for burns and infected wound and it is more effective as a dressing than many other present alternatives.

In poultry production organic acids are mainly used in order to sanitize the feed having Salmonella infection (Hinton *et al.*, 1985; Barchieri and Barrow, 1996; Thompson and Hilton, 1997). Organic acids (OA) in their undissociated forms are able to pass through the cell membrane of the bacteria, where they dissociate to produce H^+ ions which lower the pH of bacterial cell causing the organism to use its energy to restore the normal balance. Whereas the $RCOO^-$ ions, produced from the acid can disrupt DNA, hampering protein synthesis and putting the organism in stress. As a result the organism cannot multiply rapidly (Nursey, 1997).

Prebiotics are non-digestible feed ingredients that beneficially affect the host by selectively stimulating the growth or activity of one or a limited number of bacterial species, already resident in colon and thus attempt to improve host health (Gibson and Roberfroid, 1995; Ganguly & Mukhopadhayay, 2011). Mainly prebiotics are small fragments of carbohydrates and commercially available as oligosaccharides of galactose, fructose or mannose.

Among these, mannan oligosaccharide obtained from *Saccharomyces* spp. of yeast outer cell wall maintain gut health by adsorption of pathogenic bacteria, containing type-I fimbriae or by agglutinating different bacterial strains (Spring *et al.*, 2000) and increase villi length uniformity & integrity (Loddi *et al.*, 2004). Paul *et al.* (2012) revealed the effect of yeast cell wall preparation (of *Saccharomyces cerevisiae* origin) as an immunomodulator of the innate immune response. To evaluate its effect on chicken, yeast cell wall preparation (Nutriferm™) was administrated orally to 1 week old broilers @ 0.4gm and 0.8gm per kg feed for 15 days and then switched back to control diet for 20 days. Similar number of birds was kept separately as control with normal feeding regime. Non-specific immunity was assessed in randomly selected treated and control birds at the end of the experimental feeding period ('0' day) by performing neutrophil, monocyte and lymphocyte functional assay *in vitro*. In both dose group super oxide anion production by neutrophil was increased gradually up to '10' day post treatment (DPT) and then decreased to '20' DPT. The increase was significant ($P<0.5$) compared to control birds. In vitro nitrite production by monocyte was found to be high in treated birds than control. In 0.4gm treatment group in vitro non-specific lymphocyte proliferation and IL-2 production was first increased and then decreased abruptly. But in 0.8gm treatment group in vitro non-specific lymphocyte proliferation and IL-2 production was increased and then decreased gradually. The findings of the study showed that 15 days oral administration of yeast cell wall preparation on both the doses improve innate immune responses in the broiler chicks.

Effects of buffered propionic acid in presence and absence of bacitracin or roxarsone were reported earlier (Izat *et al.*, 1990) in which significant increase in dressing percentage for female broilers and a significant reduction in abdominal fat of males at 49 days (Versteegh and Jongbloed, 1999) tested the effect of dietary lactic acid on performance of broilers from 0 to 6 weeks age. Body weight gain tended to be greater, whereas feed to gain rations were significantly improved where birds were fed 2% lactic acid as prebiotic. Beneficial effects of different organic acids like formic acid, propionic acid, lactic acid ammonium formate and calcium propionate etc. as growth promoter and prebiotic have been studied earlier for having substitute to antibiotic.

Paul *et al.* (2012) studied the effect of purified α-glucan from an edible mushroom (*Pleuratus florida*) as an immunomodulator on the innate immune responses in broiler chicken (40 No.), purified mushroom glucan was administered orally to 1 week old broiler chicks (20 No.) @ 20 mg/kg feed for 15 days and then switch back to control diet. Similar number of birds was kept separately as control with normal feeding regime. Non-specific immunity and protective ability were assessed in treated and control birds at the end of the experimental feeding period (0 day) by performing neutrophil, macrophage and lymphocyte functional assay in vitro and challenged with virulent field isolate of Newcastle Disease. Superoxide anion production by neutrophil, in vitro non-specific lymphoproliferation and IL-2 production were increased gradually up to 10 days post treatment and the increase was highly significant ($P<0.05$) compared to control birds. In vitro nitrite production by macrophage was found to be high in treated birds. Also mushroom glucan as a feed supplement significantly provided protection against Newcastle Disease. The result showed the potentiality of α-glucan (mushroom origin) as an immunostimulant in poultry.

Immunomodulation Activity of the Prebiotics

An immunomodulator is a substance (*e.g.* a drug) which has an effect on the immune system. Different Immunomodulators stimulate the immune system. Biological activities of immunomodulator are influenced by different physicochemical parameters, such as solubility, primary structure, molecular weight, branching and polymer charge (Bohn and BeMiller, 1995). During the development of immune reactions, immunomodulating the effects of α-glucans have been well established (Vetvicka and Sima, 2004).

Immunostimulants stimulate the immune system. The term immunostimulant can be used interchangeably with immunomodulator, adjuvant and biological response modifier. Immunostimulators can be in the form of drugs and nutrients. They stimulate the monocyte-macrophage system and thereby modulate the immune system of the body. *Lactobacillus* spp., *Streptomyces* spp., *Aspergillus* spp. etc. are the organisms that can be used as immunostimulants (Ganguly *et al.*, 2010).

There are two main categories of immunostimulators: a) Specific immunostimulators provide antigenic specificity in immune response, such as vaccines or any antigen. For specific immune response, hosts should have prior exposure to an antigen after which recognition and subsequent activation occurs through a coordinated action of B-lymphocytes and T-cells. B cells are lymphocytes that play a large role in the humoral immune response (as opposed to the cell-mediated immune response, which is governed by T cells. b) Non-specific immunostimulators act irrespectivly of antigenic specificity to enhance the immune response of other antigens or to stimulate components of the immune system without antigenic specificity, such as glucans, synthetic drug levamisole etc. Many endogenous substances are non-specific immunostimulators. For example, Glucans and mannans possess non-specific immunostimulatory effect.α-glucan is a polymer of glucose consisting of linear backbone of α-1,3 linked D-glucopyranosyl residues with varying degree of branching from the C_6 position (Bohn and Bemiller, 1995). α-glucans are major components of yeasts, mushrooms and fungal mycelia. Mannan is a plant polysaccharide that is a polymer of the sugar mannose. Detection of mannan leads to lysis in the mannan binding lectin pathway. Immunostimulants as a feed additive significantly provides protection against pathogen and upregulates phagocytosis, bacterial killing and oxidative burst.

Mode of Immunomodulation

Probiotic microorganisms in the gut stimulate the immune response of host system in within two ways (De *et al.*, 2009). They can migrate through the gut wall as viable cells thereby multiplying to a limited extent or the released antigens by the dead organisms are absorbed and stimulate the immune response directly. Probiotics generally find their applications in poultry feed because of their positive effects growth rate, improved feed conversion and improved resistance to diseases. Probiotics have a positive effect on host immune response through increased activity of macrophages shown by enhanced ability to phagocytose organisms or carbon particles, increased production of systemic antibody *e.g.* IgM and interferon and increased effect of local antibody at mucosal surfaces such as gut wall. The effect of probiotics on the host immune system can be measured by estimating the levels of macrophage enzymes.

Prebiotics have the potentiality to enhance many host biological responses and reduce the mortality of fishes caused by invasion by pathogens. However, the anaerobic intestinal tract microbiota of commercially important fishes, such as channel catfish, hybrid striped bass, tilapia and salmonids need to be investigated to determine if there are any particular bacterial species to be enhanced with the use of prebiotics. By increasing the production of volatile fatty acids (VFAs) in the gastrointestinal (GI) tract, hosts benefit by recovering some of the lost energy from indigestible dietary constituents and by inhibiting potential pathogenic bacteria (Manning and Gibson, 2004; Vazquez *et al.*, 2005). The VFAs produced are also indicative of the microbial population present in the GI tract (Nisbet, 2002).

Prebiotics have many beneficial effects such as increased disease resistance and improved nutrient availability (De *et al.*, 2009). As such, prebiotics they have the potential to increase the efficiency and sustainability of livestock and poultry production.

The most commonly used prebiotic preparations are fructooligosaccharide (FOS), transgalactooligosaccharide (TOS), inulin, glucooligosaccharide, xylooligosaccharide, isomaltooligosaccharide, soybeanoligosaccharide, polydextrose, lactosucrose (Vulevic *et al.*, 2004; Propulla, 2008). Natural sources of prebiotics in vertebrates include chicory, onion, garlic, leek, tomato, honey etc.

Application of Prebiotics as Imunostimulants

Glucans having a strong immunomodulating activity have been well studied in fishes and livestock (Anderson, 1992). Only a few immunostimulants can be used for commercial farming purposes (Siwicki *et al.*, 1998). Many studies have been carried out to measure the effects of glucans on host immunity. Some investigators have adopted the in vitro culture of macrophages with glucan (Cook *et al.*, 2001), but in vivo studies have been carried out by the majority of workers (Sahoo and Mukherjee, 2001; Ortuno *et al.*, 2002). Under intensive conditions, individuals are more susceptible to microbial infections especially in their growing stages (Smith et al., 2003). During stress, immunostimulants can provide resistance to pathogens. There are two types of glucans á- and α- glucans the numbers of which clarify the type of O-glycosidic bond. Glucans are commercially significant as immunostimulating agents. Different types of α-glucans have been used successfully to increase resistance of poultry, fish and crustaceans against bacterial and viral infections (Paulsen *et al.*, 2001; Bagni *et al.*, 2005; Ganguly *et al.*, 2009; Ganguly & Mukhopadhayay, 2011; Paul *et al.*, 2012). It has been seen that health, growth and general performance of livestock and poultry may be improved by the use of α-glucans (Paul *et al.*, 2012). Product source, animal species, development stage of the target organism, dose and type of glucan, route, time schedule of administration and the association with other immunostimulants affect the immunomodulatory effects of glucans (Guselle *et al.*, 2007).

The immunostimulatory effects of glucan, chitin, lactoferrin, levamisole, vitamins B and C, growth hormone and prolactin have been reported. These immunostimulants mainly facilitate the function of phagocytic cells and increase their bactericidal activities. Several immunostimulants also stimulate the natural killer cells, complement, lysozyme and antibody responses of host. The most effective method of administration of immunostimulants is by injection. Oral and immersion methods have also been reported, but the efficacy of these methods decreases with long-term administration. Overdoses of several immunostimulants induce immunosuppression. Growth promoting activity has been noted in individuals treated with glucan or lactoferrin. Immunostimulants can overcome immune suppression by sex hormones (Ganguly *et al.*, 2009; 2010).

For the effective use of immunostimulants, the timing, dosages, method of administration and the physiological condition of the host needs consideration. Immunostimulants can reduce the losses caused by disease in commercial poultry rearing; however, they may not be effective against all diseases (Ganguly *et al.*, 2009; 2010).

Effect on Live Body Weight Gain

Eidelsburger and Kirchgessner (1994) reported that calcium formate alone or in combination with other acids when given at the rate of 0.5% and 1.5%, increased FCR and growth performance in broiler chickens up to 35 days of age. Benedetto (2003) also observed mix of organic acids (ACIDLAC) used as a replacer of AGPs and improved production performance in breeding hens along with other beneficial effects. Mairoka *et al*. (2004) also reported that mixture of organic acids, as a substituted of AGPs improved the performance of birds even in absence of antibiotic. Savage *et al*. (1997) concluded from a dose responsive study (0-0.33%) that MOS @ 0.11%, maximized weight gain in poultry up to 0-8 weeks of age. Stanley *et al*. (2000) found same type of effect with supplementation of 0.1% MOS on hybrid Tom's body weight gain. Parks *et al*. (2001) reported from a study from turkeys supplemented with MOS that MOS may be utilized as a alternative to AGPs to improve turkey performance.

It has already been reported that 1% formic acid or 1.45% calcium formate did not affect live weight of broiler chicken (Izat *et al*., 1990). It was found out that 80% formic acid and 20% propionic acid mixture added at 1% level to broiler chicken ration did not affect live weight (Kaniawati *et al*., 1992). It has also been reported that formic acid and propionic acid mixture (85% and 15%) added at 1% level to the broiler chicken ration in the initial period did not affect weight gain (Visek, 1978). Reports have also been made about significant increased in body weight gain with the supplementation of 0.5% lactic acid in drinking water (Veeramani *et al*., 2003). It was also revealed that increased in body weight with supplementation of lactic acid. The mix of organic acids improves performance of birds (Maiorka *et al*., 2004). From a dose responsive study it was concluded (0-0.33%) that MOS @ 0.11%, maximized weight gain in poultry up to 0-8 weeks of age. The same type of effect was found with supplementation of 0.1% MOS on hybrid Tom's body weight gain (Parks *et al*., 2001) conducted in turkeys supplemented with MOS that MOS may be utilized as a alternative to AGPs to improve turkey performance (Podolsky, 1995; Pelicano, 2003).

Dietary supplementation of different feed additives *e.g.* immunostimulants, probiotics and prebiotics usually in small quantities for the purpose of fortifying it with certain nutrients have been found to be beneficial for improving immune status, feed efficiency and growth performance of living beings also. Ganguly *et al*. (2012) highlighted the effect of dietary prebiotics on different body growth parameters of fish. However, the application of prebiotics in fish feed is now gradually gaining importance in commercial livestock farming practices, the article stresses on the effect of prebiotics on live body weight gain, dressing percentage, weight of vital organs and muscles and mean villus lengths in digestive tract of fish along with their application as growth promoters in commercial poultry and aquafeed (Gatesoupe, 2005, Ganguly *et al*., 2010).

Effect On Dressing Percentage and Weight of Vital Organs and Muscles

Pelicano (2005) reported that there was effect on the dressing percentage and weight of different organs and muscles at 21 and 42 days without any major influence on the dressing percentage, organ and muscle weight under different treatment groups with organic acid salts individually and its combination in broiler birds. Higher villus height in duodenum, jejunum in small intestine was reported with most organic acidifier in diet of broiler (Loddi *et al*., 2004). Again it was reported higher villus height in the ileum with the diet based on organic acidifier compared with diet fed without MOS + organic acidifier (Savage *et al.*, 1997). Therefore, the supplementation of organic acidifier may increase villus height of different parts of small intestine. So organic acidifier reduces the growth of many pathogenic and non-pathogenic intestinal bacteria, decreases intestinal colonization and reduces infections process, ultimately decreasing inflammatory process at the intestinal mucosa. It increases villus height and function of secretion, digestion and absorption of nutrients can be appropriately performed by the mucosa (Iji and Tivey, 1998). It was also reported positive effects of the use of prebiotics on the intestinal mucosa among which a significant increase in villus height of three segments of small intestine of birds, age one week and supplemented with MOS (Maiorka *et al.*, 2004). In the present study, use of organic acidifier in diet significantly increased villus height of different segment of small intestine of broiler possibly by reducing intestinal colonization of pathogenic and non-pathogenic bacteria, which is complied in the findings of Savage *et al.* (1997) and Loddi *et al.* (2004) respectively.

Effect on Increase in Villi Length

Loddi (2003) described higher villi in the intestinal mucosa (duodenum) of birds fed with MOS at 7 and 21 days of age respectively. Pelicano *et al.* (2005) reported that in jejunum MOS + OA resulted in higher villi in the jejunum (P<0.01) followed by the diets containing MOS based prebiotics While In case of ileum present investigation was partially corroborated with the findings of Pelicano *et al.* (2005) who observed higher villi length when birds were fed with prebiotic based on MOS, compared to the control group.

Microorganisms that is sensitive to acid pH and results in higher villi length (Radecki and Yokoyama, 1991). Some bacteria may recognize binding sites on the prebiotics instead of intestinal mucosa and the colonization by pathogenic bacteria in intestine is thus reduced. Therefore, besides a lower infection incidence, there is an increase in the absorption of available nutrients, a mechanism that directly affects the recovery of the intestinal mucosa, increasing villi length. These results disagree to those obtained by Pelicano *et al.* (2003) and Santin *et al.* (2001) respectively, who found no difference in ileal villi length with the use of probiotics and prebiotics.

Roy *et al.* (2012) conducted experimental studies to evaluate the pathomorphological effect of different combination of organic acids *viz.* formic acid, propionic acid and lactic acid as a replacer of growth promoter antibiotics. The birds were divided into six equal groups *viz.* negative control (C_1), positive control (C_2) and treatment groups (T_1, T_2, T_3 and T_4). Birds of group C_1 were supplied with diet without any antibiotics or acid, group C_2 with Virginiamycin @ 500 gm/quintal feed, group T_1 with 0.3% ammonium formate group T_2 with 0.3% calcium propionate, group T_3 with 0.15% ammonium formate and 0.15% calcium propionate and group T_4 with 0.1% ammonium formate, 0.1% calcium propionate and 0.1% calcium lactate. Body weight gain was higher in C_1 and C_2 compared with treatment groups in first two weeks, but pathological changes were maximum in negative control, *i.e.* after 6 weeks weight gain was significantly better in T_3 and T_4 than in groups C_1, T_1 and T_2 groups. Pathomorphological changes in group C_1 and T_2. Group C_2 and T_1 also showed same types of changes and in group T_3 and T_4 the changes were less. The most common changes among the groups were thickening of muscular layer, accumulation of inflammatory cells and congested blood vessels. In T_1 group, rupture of villi was also seen. In T_3 and T_4 the intestinal glands were hypertrophied. Based on present investigation, it is suggested that combination of organic acid may be used in broiler feed as a replacer of growth promoter antibiotic instead of using only one acid.

It was reported that combined use of MOS and organic acid salts in poultry feed can be used as an alternative to the antimicrobial and antibiotic growth promoters to achieve good health for sustainable and economic poultry production in India (Das *et al.*, 2012).

Effect on Gut Microbial Load

The findings that MOS and OA successfully reduces bacterial load in the intestine of broiler birds were in accordance with the findings of Newman (1994), Lon (1995), Spring *et al.* (2000) and Fairchild *et al.* (2001). Stanley *et al.* (2004) concluded that yeast cell culture residue (YCR) treated broiler chicks resulted lower intestinal Coliform population in comparison to control and other antibiotic treated (lasalocid @ 90.7kg/ton, bacitracin @50gm/ton) groups. Sims *et al.* (2004) also found that MOS + BMD treated turkeys showed significantly lower *Clostridium perfringens* population in the gut than control at 6 weeks of age. MOS is believed to block type-I fimbriae and prevents pathogens from attaching to the intestinal lining and passes them out of the gut. (Dawson and Pirulescu, 1999). Bacteriological studies of intestine of broiler birds revealed reduced *E. coli* compared with the untreated control group. Salmonellae were not found in any group. No significant difference was observed among control and treatment groups regarding the number of *Clostridium* sp. present in the intestine. The above result clearly indicated that consumption of prebiotics mixed with poultry feed reduced the load of coliform bacteria in the intestine (Roy *et al.*, 2012).

It can be summarized that supplementation of poultry feed with dietary probiotics in proper proportions can enhance the immune system of the host by providing increased resistance to infections.

In livestock production organic acids are mainly used in order to sanitize the feed having *salmonella* infection (Hinton et al, 1985; Barchieri and Barrow, 1996; Thompson and Hilton, 1997). Organic acids (OA) in their undissociated forms are able to pass through the cell membrane of the bacteria, where they dissociate to produce H^+ ions which lower the pH of bacterial cell causing the organism to use its energy to restore the normal balance. Whereas the $RCOO^-$ ions, produced from the acid can disrupt DNA, hampering protein synthesis and putting the organism in stress. As a result the organism cannot multiply rapidly (Nursey, 1997).

Prebiotics are non-digestible feed ingredients that beneficially affect the host by selectively stimulating the growth or activity of one or a limited number of bacterial species, already resident in colon and thus attempt to improve host health (Gibson and Roberfroid, 1995). Mainly prebiotics are small fragments of carbohydrates and commercially available as oligosaccharides of galactose, fructose or mannose (Ganguly *et al*, 2010; 2013b,c).

Among these, mannan oligosaccharide obtained from *Saccharomyces* spp. of yeast outer cell wall maintain gut health by immunomodulation (Paul *et al*, 2013) and by adsorption of pathogenic bacteria containing type-I fimbriae or by agglutinating different bacterial strains (Spring *et al*, 2000) and increase villi length uniformity & integrity (Loddi *et al*, 2004).

Effects of buffered propionic acid in presence and absence of bacitracin or roxarsone were reported earlier (Izat *et al*, 1990) in which significant increase in dressing percentage for female broilers and a significant reduction in abdominal fat of males at 49 days (Versteegh and Jongbloed, 1999) tested the effect of dietary lactic acid on performance of broilers from 0 to 6 weeks age. Body weight gain tended to be greater, whereas feed to gain rations were significantly improved where birds were fed 2% lactic acid as prebiotic. Beneficial effects of different organic acids like formic acid, propionic acid, lactic acid ammonium formate and calcium propionate etc. as growth promoter and prebiotic have been studied earlier for having substitute to antibiotic (Ganguly *et al*, 2010; Ganguly and Mukhopadhayay, 2012).

Xylanase is the name given to a class of enzymes which degrades the linear polysaccharide beta-1, 4 xylan to xylose, thus breaking down hemicelluloses which are a major component of the cell wall of the plants. Xylanases are known to increase protein digestibility of wheat and this is attributed particularly to release of protein from the xylan enriched aleurone layer. Xylanase supplementation improves conjugated bile acid function in intestinal contents and increase villus size of small intestine wall in broiler (Bar *et al*, 2012; Ganguly, 2013b,c). Supplementing broiler diets with combination of xylanase and α-glucanase improves the nutritive value of the diet (Veldman and Vahl, 1994). The addition of Xylanase improves weight gain, feed intake, feed efficiency, AME and decreased water intake (Wu *et al*, 2004) and Vitamin E content of liver in broiler was significantly improved by addition of xylanase (Danicke et al, 1999; 2001). Nutri-xylanase is a bacterial xylanase processed from *Bacillus subtilis* and produced by a micro-filtration advanced fermentation technique.

The plant derived and herbal growth promoters supplemented in the diet or added in the drinking water in the broiler and poultry birds have a promising biological effect on their growth performance, to reduce the pathogenic bacteriological load in different parts of digestive tract and to increase villus height in different segments of small intestine mainly in duodenum. The plant derived growth promoter enhance productive performance of the broiler in terms of body weight gain with minimum alteration of gut morphology and the possibility of bacterial invasion can be regulated (Ganguly, 2013b;c).

Mairoka *et al* (2004) reported that mixture of organic acids, as a substituted of AGPs improved the performance of birds even in absence of antibiotic. Savage et al (1997) concluded from a dose responsive study (0-0.33%) that MOS @ 0.11%, maximized weight gain in animals up to 0-8 weeks of age. Stanley et al (2000) found same type of effect with supplementation of 0.1% MOS on hybrid Tom's body weight gain. Eidelsburger and Kirchgessner (1994) reported that calcium formate alone or in combination with other acids when given at the rate of 0.5% and 1.5%, increased FCR and growth performance in broiler chickens up to 35 days of age. Benedetto (2003) also observed mix of organic acids (ACIDLAC) used as a replacer of AGPs and improved production performance in breeding hens along with other beneficial effects. Parks *et al* (2001) reported from a study from turkeys supplemented with MOS that MOS may be utilized as a alternative to AGPs to improve turkey performance.

It has also been reported that formic acid and propionic acid mixture (85% and 15%) added at 1% level to the broiler chicken ration in the initial period did not affect weight gain (Visek, 1978). Reports have also been made about significant increased in body weight gain with the supplementation of 0.5% lactic acid in drinking water (Veeramani *et al*, 2003). The mix of organic acids improves performance of birds (Maiorka *et al*, 2004). From a dose responsive study, it was concluded (0-0.33%) that MOS @ 0.11%, maximized weight gain in animals up to 0-8 weeks of age. The same type of effect was found with supplementation of 0.1% MOS on hybrid Tom's body weight gain (Parks *et al*, 2001) conducted in turkeys supplemented with MOS that MOS may be utilized as a alternative to AGPs to improve turkey performance (Podolsky, 1995; Pelicano, 2003). It has already been reported that 1% formic acid or 1.45% calcium formate did not affect live weight of broiler chicken (Izat *et al*, 1990). It was found out that 80% formic acid and 20% propionic acid mixture added at 1% level to broiler chicken ration did not affect live weight (Kaniawati *et al*, 1992). It was also revealed that increased in body weight with supplementation of lactic acid.

A series of experiments was conducted by Paul *et al*. (2013) to evaluate the various aspects and effects of different combination of organic acids *viz.*, formic acid and propionic acid as a replacer of growth promoter antibiotic(s) in ducks. The ducks were divided into five equal groups with one as Control. Studies on body weight gain revealed that after 48 weeks body weight gain was higher in treated groups as compared to the control Group C (control).

It has been reported that higher villus height in the ileum with the diet based on organic acidifier compared with diet fed without MOS + organic acidifier (Savage *et al*, 1997). Dressing percentage and weight of different organs and muscles at 21 and 42 days there was no major influence on the dressing percentage, organ and muscle weight under different treatment groups with organic acid salts individually and its combination in broiler birds. These findings in line of earlier report (Pelicano, 2005). Higher villus height in duodenum, jejunum in small intestine was reported with most organic acidifier in diet of broiler (Loddi *et al*, 2004). The supplementation of organic acidifier may increase villus height of different parts of small intestine. So, organic acidifier reduces the growth of many pathogenic and non-pathogenic intestinal bacteria, decreases intestinal colonization and reduces infections process, ultimately decreasing inflammatory process at the intestinal mucosa. It increases villus height and function of secretion, digestion and absorption of nutrients can be appropriately performed by the mucosa (Iji and Tivey, 1998). The positive effects of the use of prebiotics on the intestinal mucosa with significant increase in villus height of three segments of small intestine of birds supplemented with MOS is also reported (Maiorka *et al*, 2004).

Silverslides and Bedford (1999) and Bar *et al*, (2012) showed xylanase supplementation had a positive body weight gain and the feed to gain ration. Danicke et al (2001) found addition of xylanase significantly increased the weight gain up to 21 days of age and decreased the feed to gain ratio slightly. Mathlouthi *et al*, (2003) showed feed efficacy and body weight gain was improved with the supplementation of xylanase. Wu *et al*, (2004) observed that the xylanase supplementation significantly improved weight gain, feed efficiency and AME. Wu *et al*, (2004) observed that addition of xylanase and phytase reduced the relative weight gain of the small intestine by 15.5% and 11.4% respectively. Yubo *et al*, (2005) reported body weight and feed per gain ratio was improved (P<0.05) by xylanase supplementation in the first 2-3 weeks in broilers. Ahmad *et al*, (2007) noticed xylanase supplementation increased body weight, feed intake and feed gain ratio. Liu *et al*, (2007) observed xylanase supplementation increased body weight gain from 0-21days of age of broilers. Gao *et al*, (2008) found that the supplementation of xylanase enzyme improved (P<0.05) growth performance and feed conversion efficiency (FCE). Mannio (1981) reported that the body weight gain was improved by 12 to 25% and feed consumption was increased by 3 to 21% when chicks at 4 weeks of age fed diets supplemented with enzyme like xylanase. Veldman and Vahl (1994) noticed xylanase supplementation improved food conversion ratio by 2.2-2.9% and body weight gain by 0.2-2.5%.

Das *et al*, (2012) reported that MOS (mannan oligosaccharide) and organic acid treated groups of Japanese quails (*Coturnix Coturnix Japonica*) produced consistently higher villi length in treated birds and MOS in poultry feed can be used as alternatives to the antimicrobial and antibiotic growth promoters and can be used to achieve good health for sustainable and economic poultry production. On the other hand, experimental studies were conducted by Roy *et al*, (2012) to evaluate the pathomorphological effect of different combination of organic acids *viz.*, formic acid, propionic acid and lactic acid as a replacer of growth promoter antibiotics in poultry birds. The birds were divided into six equal groups of negative control (C1), positive control (C2) and four treatment groups. Birds of group C1 were supplied with diet without any antibiotics or acid, group C2 with Virginiamycin @ 500 gm/100kg feed, group T1 with 0.3% ammonium formate group T2 with 0.3% calcium propionate, group T3 with 0.15% ammonium formate and 0.15% calcium propionate and group T4 with 0.1% ammonium formate, 0.1% calcium propionate and 0.1% calcium lactate. Body weight gain was higher in C1 and C2 compared with treatment groups in first two weeks, but pathological changes were maximum in negative control, *i.e.* after 6 weeks, weight gain was significantly better in T3 and T4 than in groups C1, T1 and T2 groups. Bacteriologically, significant (P<0.01) reduction of *E. coli* in T1 and T4 was evident. Pathomorphological changes in group C1 and T2 were maximum. Group C2 and T1 showed same types of changes but the changes were less severe in group T3 and T4. The most common changes among the groups were thickening of muscular layer, accumulation of inflammatory cells and congested blood vessels. Based on present investigation, it is suggested that combination of organic acid may be used in broiler feed as a replacer of growth promoter antibiotic instead of using only one acid. Pelicano *et al* (2005) observed higher villi length in ileal region when birds were fed with prebiotic based on MOS, compared to the control group. Loddi (2003) described higher villi in the intestinal mucosa (duodenum) of birds fed with MOS at 7 and 21 days of age respectively. Pelicano *et al* (2005) reported that in jejunum MOS + OA resulted in significantly higher villi in the jejunum (p<0.01) followed by the diets containing MOS based prebiotics. Microorganisms that is sensitive to acid pH and results in higher villi length (Radecki and Yokoyama, 1991). Some bacteria may recognize binding sites on the prebiotics instead of intestinal mucosa and the colonization by pathogenic bacteria in intestine is thus reduced. Therefore, besides a lower infection incidence, there is an increase in the absorption of available nutrients, a mechanism that directly affects the recovery of the intestinal mucosa, increasing villi length. These results disagree to those obtained by Pelicano *et al* (2003) and Santin *et al* (2001) respectively, who found no difference in ileal villi length with the use of probiotics and prebiotics.

The bacteriological studies of different portions of small intestine revealed that total coliform count and *Clostridium perfringens* count ($\log_{10}$ CFU/g) was significantly (P<0.05) reduced in the small intestine of the groups of birds orally administered with yeast glucan as compared to the untreated control group. Salmonella sp. was not found in any group. No significant results of Lactobacillus count ($\log_{10}$ CFU/g) were noticed in the intestinal digesta of the ducks in treated groups. Study on villus height of different potions of small intestine (i.e. duodenum, jejunum and ileum) revealed significantly higher villus height in treated groups as compared to the control group [Paul *et al*, 2013].

Hrangkhawl et al. (2013) conducted series of experiments to study the effect of mannan oligosaccharide (MOS) and dietary organic acid supplements on body weight of broiler birds. The present investigation showed better growth performance in combination with organic acid salts in terms of body weight. It was found that mean villus length increased significantly (P<0.01) in the treatment groups rather than the control birds.

The organic acid salts and MOS can be used as alternatives to growth promoters but their combination strategy can be used to achieve good health and growth performance. The live body weight gain is better in organic acid and MOS supplemented animals. MOS and organic acids individually or in combination reduce gut microbial load and improve growth performance of animals.

References

Anderson, D.P., 1992. Immunostimulants, adjuvants and vaccine carriers in fish: applications to aquaculture. *Annu. Rev. Fish Dis.*, 2: 281-307.

Agarwal S S, Singh V K. 1969. Immnomodulators: a review of studies on Indian Medicinal Plants and Synthetic Peptides. *PINSA.* 65 (3-4): 179-204.

Bishavi B, Roychowdhury S, Ghosh S and Sengupta M. 2002. Hepatoprotective and immunomodulatory properties of *Tinospora cordifolia* in CCl_4 in toxicated mature albino rats *J. Toxicol. Sci.* 27(3): 139-146.

Chopra R N, Chopra L C, Handa K D and Kapur L D. 1982. *Indigenous Drugs of India.* 2nd edn. Dhur & Sons Pvt Ltd, Calcutta, India.

Diwanay S, Chitre D and Patwardhan 2004. Immunoprotection by botanical drugs in cancer chemotherapy. *J. Ethnopharmacol.* 90(1): 49-55.

Hellar E.D. 1975. *Res. Vet. Sci.*18: 117 (Cited by Rao *et al.* 1987). Resistance of maternal antibodies against Newcastle disease virus in chicks from immune parents and its effect on vaccination. *Indian J. Comp. Microbiol. Immunol. Inf. Dis.* 8(3): 106-110.

Kalita D N and Dutta G N. 1999. Immunomodulatory effect of levamisole upon Newcastle disease, pigeon pox and Mark's disease vaccination in broiler chicks. *Indian Vet. J.* 76: 490-492.

Karnataka B C, Shukla S K, Kumar M and Dixit V P. 1993. Immunomodulatory effect of levamisole on the antibody response to RD vaccination. *Indian J. Vet. Med.* 13(2): 48-51.

Kolte A Y, Siddiqui M F and Mode S G. 2007. Immunomodulating effect of *Withania somnifera* and *Tinospora cordifolia* in broiler birds.

Krishnamohan A V, Reddy D B, Sarma B and John Kirubharan J. 1997. Studies on the effects of levamisole against Newcastle disease virus in chicken. *Indian J. Comp. Microbiol. Immunol. Infct. Dis.* 8: 1-6.

Kujur R T. 2001. Evaluation of certain immunomodulatory agents in countering immunosuppressive effects of vaccine strain of infectious bursal disease virus in chicks. *M.V.Sc. thesis.* Rajendra Agricultural Univ., Bihar, India.

Kumar P. 2003. Studies on comparative immunomodulatory effect of herbal preparation and Vitamin E-Se in comparison to Levamisole in broiler chicks. *M.V.Sc. thesis.* Birsa Agricultural Univ., Ranchi, India.

Kuttan G and Kuttan R. 1992. Immunomodulatory activity of a peptide isolated from *Viscum album* extract. *Immunol. Invest.* 21: 285-296.

Manjrekar P N, Jolly C I and Narayan S. 1999. Comparative studies of the immunomodulatory acivities of *Tinospora cordifolia* and *Tinospora sinensis*. *Fitoterapia.* 71: 254-257.

Muruganandan S, Garg H, Lal J, Chandra S and Kumar D. 2001. Studies on the immunostimulant and anti-hepatotoxic activities of *Asparagus recemosus* root extract. *J. Med. Arom. Pl. Sci.* 22: 49-52.

Nadkarni A.K. 1954. *Indian Materia Medica*, Bombay, Popular Book Depot, 3rd edn., 1: 153-155.

Panda S K and Rao A T. 1994. Effect of levamisole on chicken infected with infectious bursal disease (IBD) virus. *Indian Vet. J.* 71: 427-439.

Rege N.N and Dahanukar S.A. 1993. Quantitation of microbicidal activity of mononuclear phagocytes : an *in vitro* technique. *J. Postgrad. Med.* 39(1): 22-25.

Rege N N, Nazareth H M, Bapat R D and Dhanukar S A. 1989. Modulation of immunosuppression in obstructive jaundice by *Tinospora cordifolia*. *Ind. J. Med. Res.* 90: 478-483.

Renoux and Renoux. 1971. Mechanism of action of some immunomodulatory drugs used in Veterinary Medicine. *Adv. Vet. Sci. Comp. Med.* 35: 43-49. 293

Soppi E, Lassila O, Vilijanen M K, Lehtonon O P and Eskole J. 1979. *Clin. Exp. Immunol.* 38: 609. (Cited by Chakraborty D and Chatterjee A. 1998. Studies on immunomodulatory effect of Levamisole in Newcastle disease vaccinated chicks. *Indian J. Comp. Microbiol. Immunol. Infect. Dis.* 19: 85-87).

Thatte U M and Dahanukar S A. 1986. Ayurveda and contemporary scientific thought. *Trends in Pharmacol. Sci.* 17: 248-257.

Thatte U M and Dahanukar S A. 1989. Immunotherapeutic modification of experimental infection by Indian medicinal plants. *Phytothe. Res.* 3: 43-49.

Vyas G P, Dholakia P M and Kathiria L G. 1987. Studies on immunomodulation by levamisole along with vaccination in chicks against Ranikhet disease. *Indian Vet. J.* 64: 456-462.

Bachère, E., 2003. Anti-infectious immune effectors in marine invertebrates: potential tools for disease control in larviculture. *Aquaculture*, 227: 427–38.

Bagni, M., Romano, N., Finoia, M.G., Abelli, L., Scapigliati, G. and Tiscar, P.G., 2005. Shortand long-term effects of a dietary yeast α-glucan (MacroGard) and alginic acid (Ergosan) preparation on immune response in sea bass (*Dicentrarchus labrax*). *Fish Shellfish Immunol.*, 18(4): 311-325.

Bohn J.A. and BeMiller J.N., 1995. *(1ÀÛÆÜ3)*-α-glucans as biological response modifiers: a review of structure-functional activity relationships. Carbohydrate polymers 28: 3-14.

Brown, G.D. and Gordon, S., 2005. Immune recognition of fungal glucans. *Cell. Microbiol.*, 7: 471-479.

Choat, J. and Clements, K., 1998. Vertebrate herbivores in marine and terrestrial environments: A nutritional ecology perspective. *Annu. Rev. Ecol. Sys.*, 29: 375-403.

Cook, M.T., Hayball, P.J., Hutchinson W., Nowak, B.F. and Hayball, J.D., 2001. Administration of a commercial immunostimulant preparation, EcoActiva™ as a feed supplement enhances macrophage respiratory burst and the growth rate of snapper, *Pagrus auratus*, Sparidae (Bloch and Schneider) in winter. *Fish Shellfish Immunol.*, 14: 333-345.

Fuller, R., 1989. Probiotics in man and animals. *J. Appl. Bacteriol.*, 66: 365-378.

Bagni, M., Romano, N., Finoia, M.G., Abelli, L., Scapigliati, G., Tiscar, P.G., *et al.* 2005. *Fish Shellfish Immunol*. 18(4): 311-25.

Brown, G.D., and Gordon, S. 2005. Cellular microbiology , 7: 471-79.

Cahill MM (1990) Bacterial flora of fishes: a review. *Microbial Ecology*. 19, 21-41.

Cho C.Y. and Kaushik S.J. 1985. Effects of protein intake on metabolizable and net energy values of fish diets. pp. 95-117, In: Nutrition and Feeding in Fish. Cowey C.B., Mackie A.M. and Bell J.G., Eds. Academic Press, London.

Cho C.Y., and, Kaushik S.J. 1990. Nutritional energetics in fish: Energy and protein utilization in rainbow trout (*Salmo gairdneri*). *World Rev. Nutr. Diet*. 61: 132-72.

Cowey C.B. 1975. Aspects of protein utilization by fish. *Proc. Nutr. Soc*. 34: 57-63

Cook, M.T., Hayball, P.J., Hutchinson, W., Nowak, B.F., and Hayball, J.D. 2003. *Fish Shellfish Immunol*. 14: 333-45.

Davies ME (1965) Cellulolytic bacteria in some ruminants and herbivores as shown by fluorescent antibody. *Journal of General Microbiology*. 39, 139–141.

De Gregorio, E., Spellman, P.T., Rubin, G.M., and Lemaitre, B. 2001. *Proc. Natl. Acad. Sci. USA.*, 98: 12590–95.

Finegold GM, Sutter VL, Mathisen GE (1983). Normal indigenous intestinal flora. In: D.J. Hentgens (ed.) Human intestinal microflora in health and disease, pp. 3-31. New York: Academic Press.

Furuichi M, Yone Y (1982). Availability of carbohydrate in nutrition of carp and red sea bream. *Bulletin of the Japanese Society for Scientific Fisheries*. 48, 945–948.

Gatlin D.M.III., Poe W.E., and Wilson R.P. 1986. Protein and energy requirements of fingerling channel catfish for maintenance and growth. J. Nutr. 116: 2121-31.

Garling D.L., Jr., and Wilson R.P. 1976. Optimum dietary protein-to-energy ratios for channel catfish fingerlings, *Ictalurus punctatus*. *J. Nutr*. 106: 1368-75.

Ganguly, S., Paul, I., and, Mukhopadhayay, S. K. 2009. Immunostimulants- Their Significance in Finfish Culture. *Fish. Chimes*. 29(7): 49-50.

Ganguly, S., Paul, I., and Mukhopadhayay, S. K. 2010a. Immunomodulatory Effects of Fungal Âeta - Glucans In Fish Farming. *Fish. Chimes*. 30(7): 64.

Ganguly, S., Paul, I., and Mukhopadhayay, S. K. 2010b. Immunostimulant, probiotic and prebiotic – their applications and effectiveness in aquaculture: A Review. *Israeli J. Aquacult.* – Bamidgeh, 62(3): 130-38.

Ganguly, S., Dora, K. C., Sarkar, S. and, Chowdhury, S. 2013a. Supplementation of prebiotics in fish feed: A Review. Rev. Fish Biol. Fisheries. 23(2): 195-99, DOI: 10.1007/s11160-012-9291-5.

Ganguly, S. 2013b. Role of proper nutrient formulation in fishery and marine sciences and related hygiene issues: A Review. *Res. J. Marine Sci*. 1(2): 17-18.

Ganguly, Subha. 2014*a*. The multivarious utilities of Neem (*Azadirachta indica*) in traditional medicine: An Exclusive Review. *Int. J. Pharm. Natural Medi*. 2(1): xx-xx. In press.

Ganguly, Subha. 2014*b*. Utility of Papaya in traditional medicine for its immense nutritional and health benefits: A Review. *J. Pharm. Biol. Res*. 2(1): xx-xx. In press.

Ganguly, Subha and Bordoloi, Ranjit. 2014. *Centella asiatica*, a potential indigenous herb of potential medicinal implication in Ayurveda and clinical therapy: A Review. *Int. J. Res. Pharm. Life Sci*. 2(1): xx-xx. In press.

Goldstein L., and Forster R.P. 1970. Nitrogen metabolism in fish. pp. 495-515. In: Comparative Biochemistry of Nitrogen Metabolism. Vol. 2. The Verteleratts, Campbell J.W., ed. Academic Press, NY.

Guselle, N.J., Markham, R.J.F., and Speare, D. J. 2007. *Journal of Fish Diseases*. 30 (2): 111-16.

Hansen GH, Strom E, Olafsen JA (1992). Effect of different holding regimes on the intestinal microflora of herring (*Clupea harengus*) larvae. *Applied and Environmental Microbiology*. 58, 461-470.·

Leahy JG, Colwell RR (1990). Microbial degradation of hydrocarbons in the environment. *Microbiological Reviews*. 54(3), 305-315.

Lesel R, Fromageot C, Lesel M (1986). Cellulose digestibility in grass carp and goldfish. *Aquaculture*. 54, 11-17.

Lindsay GJH, Harris JE (1980). Carboxymethylcellulase activity in the digestive tracts of fish. *Journal of Fish Biology*. 16, 219–233.

Meisner A, Burns J (1997). Viviparity in the Halfbeak Genera *Dermogenys* and *Nomorhamphus* (Teleostei: Hemiramphidae). *Journal of Morphology*. 234, 295–317.

Medzhitov, R., and Janeway, C.A. 1997. *Cell*. 91: 295–98.

Panigrahi, A, Kiron, V, Puangkaew J, Kobayashi T, Satoh S, Sugita, H (2005) The viability of probiotics bacteria as a factor influencing the immune response in rainbow trout *Oncorhynchus mykiss*. *Aquaculture*. 243, 241-254.

Pucci OH, Bak MA, Peressutti SR, Klein IC. Hrtig HM, Alvarez L, Wünsche (2004). Influence of crude oil contamination on the bacterial community of semiarid soils of Patagonia (Argentina).

Ringo E, Birkbeck TH (1999). Intestinal microflora of fish and fry: a review. *Aquaculture Research*. 30(2), 73-93.

Ringo E, Olsen RE (1999). The effect of diet on aerobic bacterial flora associated with intestine of Arctic charr *(Salvelinus alpinus)*. *Journal of Applied Microbiology*. 86, 22-28.

Ringo E, Strom E (1994). Microflora of Arctic charr *Salvelinus alpinus* gastrointestinal microflora of freeliving fish and effect of diet and salinity on the intestinal microflora. *Aquaculture and Fisheries Management*. 25, 623-629.

Robertsen, R., Engstad, E., and Jorgensen, J.B. 1994. α-Glucans as immunostimulants in fish. Modulators of fish immune responses. In: Stolen, J.S. and Fletcher, T.C. Editors, Models for environmental toxicology. Biomarkers, immunostimulators Vol. 1, 505 Publications, Fair Haven, NJ, pp. 83–99.

Sahoo, P.K., and Mukherjee, S.C. 2001. Fish Shellfish Immunol. 11: 683–95.

Sakata T (1990). Microflora in the digestive tract of fish and shellfish. In: Lesel R (ed.) Microbiology of Poekilotherms, New York, London: Elsevier Amsterdam, pp. 217-223.

Sakata T, Yuki T (1992). Diagnostic media for differentiation of Plesiomonas from intestinal microflora of freshwater fish. *Bulletin of Japanese Society of Scientific Fisheries*. 58(5), 977-979.

Scott P (1997). Livebearing Fishes. pp. 13. Tetra Press.

Smith LS (1989) Digestive functions in teleost fishes. In Fish Nutririon, 2nd edn. ed. Halver JE, pp 331-421, San Diego; Academic Press.

Strom E, Olafsen JA (1990). The indigenous microflora of wild-captured juvenile cod in net-pen rearing. In: Lesel R (ed.) Microbiology of Poecilotherms, pp. 181-185. New York, London: Elsevier Amsterdam.

Sugita H, Kawasahi J, Deguchi Y (1997). Production of amylase by the intestinal microflora in cultured freshwater fish. *Letters of Applied Microbiology*. 24, 105-108.

Sugita H, Tsunohara M, Ohkashi T, Deguchi Y (1988). The establishment of an intestinal microflora in developing goldfish *Carassius auratus* ponds. Microbial Ecology. 15, 333-344.

Syvokieni J (1989). Symbiotic digestion in hydrobionts and insects. *Vilnius: Mokslas*.

Takeuchi T (1991). Digestion and Nutririon of Fish. In *Fish Physiology* ed. Itazawa Y. and I. Hanyu, pp. 67-101. Tokyo: *Koseisha Koseikaku* (in Japanese).

Westerdahl A, Olsen JC, Kjelleberg S, Comway P (1991). Isolation and characterization of turbot (*Scophthalmus maimus*) associated bacteria with inhibitory effects against *Vibrio anguillarum*. *Applied and Environmental Microbiology*. 57, 2223-2228.

Vesta Skrodenyte-Arbaeiauskiene (2000). Proteolytic activity of the roach (*Rutilus rutilus* L.) intestinal microflora. *Acta Zoologica Lituanica*. 10, 3.

Voverienë G, Mickënienë L, Ðyvokienë J (2002). Hydrocarbon-degrading bacteria in the digestive tract of fish, their abundance, species composition, and activity. Acta Zoologica Lituanica. 12, 3.

Vetvicka, V., and Sima, P. 2004. ISJ , 1: 60-65.

Zekovic, D.B., and Kwiatkowski, S. 2005. Critical reviews in biotechnology, 25: 205-30.

Guselle, N.J., Markham, R.J.F., and Speare, D. J., 2007. Timing of intraperitoneal administration of â- 1,3/1,6 glucan to rainbow trout, *Oncorhynchus mykiss* (Walbaum), affects protection against the microsporidian Loma salmonae. *J. Fish Dis.*, 30 (2): 111–116.

Hagi, T., Tanaka, D., Iwamura, Y. and Hosino, T., 2004. Diversity and seasonal changes in lactic acid bacteria in the intestinal tract of cultured freshwater fish. *Aquaculture*, 234: 335-346.

Li, P. and Gatlin, D. M. III, 2004. Dietary brewer's yeast and the prebiotic Grobiotic TM AE influence growth performance, immune response and resistance of hybrid striped bass (*Morone chrysops* X *M. saxatilis*) to *Streptococcus iniae* infection. *Aquaculture*, 231: 445-456.

Li, P. and Gatlin, D.M. III., 2005. Evaluation of prebiotics GroBiotic®-A and brewer's yeast as dietary supplements for sub-adult hybrid striped bass (*Morone chrysops* X *M. saxatilis*) challenged in situ with *Mycobacterium marinum*. *Aquaculture*, 248:197-205.

Manning, T. and Gibson, G. 2004. Prebiotics. *Best Practice & Research in Clinical Gastroenterology*, 18: 287-298.

Nikoskelainen, S., Ouwehand, A.C., Bylund, G., Salminen, S. and Lilius, E., 2003. Immune enhancement in rainbow trout (*Oncorhynchus mykiss*) by potential probiotics bacteria (*Lactobacillus rhamnosus*). *Fish Shellfish Immunol.*, 15: 443-452.

Nisbet, D., 2002. Defined competitive exclusion cultures in the prevention of enteropathogen colonization in poultry and swine. *Antoine van Leeuwenhoek*, 81: 481-486.

Ortuno, J., Cuesta, A., Rodriguez, A., Esteban, M.A., and Meseguer, J., 2002. Oral administration of yeast, *Saccharomyces cerevisiae*, enhances the cellular innate immune response of gilthead seabream (*Sparus aurata* L.). *Vet. Immunol. Immunopath.*, 85: 41– 50.

Panigrahi, A., Kiron, V., Puangkaew, J., Kobayashi, T., Satoh, S. and Sugita, H., 2005. The viability of probiotics bacteria as a factor influencing the immune response in rainbow trout Oncorhynchus mykiss. *Aquaculture*, 243: 241-254.

Parker, R.B., 1974. Probiotics, the other half of the antibiotic story. *Anim. Nutr. Health,* 29: 4-8.

Paulsen, S.M., Engstad, R.E. and Robertsen, B., 2001. Enhanced lysozyme production in Atlantic salmon (*Salmo salar* L.) macrophages treated with yeast â-glucan and bacterial lipopolysaccharide. *Fish Shellfish Immunol.,* 11(1): 23-37.

Propulla, S.G., 2008. Prebiotic: an emerging functional food. Compendium of Winter School on "Harnessing microbial diversity for use in animal nutrition", 4-24 Nov., 2008. National Institute of Animal Nutrition and Physiology, Adugodi, Bangalore-560030, India, pp. 47-53.

Ringo, E., 1993. Does dietary linoleic acid affect intestinal microflora in Arctic Charr, *Salvelinus alpinus* (L.)? *Aquaculture and Fisheries Management,* 24: 133-135.

Ringo, E. and Gatesoupe, F.J., 1998. Lactic acid bacteria in fish: a review. *Aquaculture,* 160: 177-203.

Ringo, E. and Olsen, R.E., 1999. The effect of diet on aerobic bacterial flora associated with intestine of Arctic Charr, *Salvelinus alpinus* (L.) *J. Appl. Microbiol.,* 88: 22-28.

Robertsen, R., Engstad, E., and Jorgensen, J.B., 1994. â-Glucans as immunostimulants in fish. Modulators of fish immune responses. In: J.S. Stolen and T.C. Fletcher, Editors, Models for environmental toxicology. Biomarkers, Immunostimulators Vol. 1, 505 Publications, Fair Haven, NJ, pp. 83–99.

Sahoo, P.K. and Mukherjee, S.C., 2001. Effect of dietary â-1, 3 glucan on immune responses and disease resistance of healthy and aflatoxin B1-induced immunocompromised rohu (*Labeo rohita* Hamilton). *Fish Shellfish Immunol.,* 11: 683–695.

Salinas, I., Cuesta, A., Esteban, Angeles and Meseguer, J., 2005. Dietary administration of *Lactobacillus dulbrueckii* and *Bacillus subtilis*, single or combined, on gilthead seabream cellular innate immune responses. *Fish Shellfish Immunol.,* 19: 67-77.

Siwicki, A.K., Morand, M., Terech-Majevska, E., Niemczuk, W., Kazun, K. and Glabsky, E., 1998. Influence of immunostimulant on the effectiveness of vaccines in fish: in vitro and in vivo study. *J. Appl. Ichthyol.,* 14: 225–227.

Smith, V., Brown, J. and Hauton, C., 2003. Immunostimulation in crustaceans: does it really protect against infection. *Fish Shellfish Immunol.,* 15: 71–90.

Titus, E. and Ahearn, G., 1988. Short-chain fatty acid transport in the intestine of a herbivorous teleost. *J. Exp. Biol.,* 13: 577-94.

Vazquez, J.A., Gonalez, M.P. and Murado, M.A., 2005. Effects of lactic acid bacteria cultures on pathogenic microbiota from fish. *Aquaculture,* 245: 149-161.

Vulevic, J., Rastall, R.A., and Gibson, G.R., 2004. Developing a quantitative approach for determining the invitro prebiotics potential of dietary oligosaccharides. *FEMS Microbiology Letters,* 236: 153-159.

Bar N, Mukhopadhayay SK, Ganguly S, Pradhan S, Patra NC, Pal S, Goswami J, Singh YD, Halder S (2012). Study on probiotic effect of xylanase supplementation in broiler feed. *Indian J. Anim. Nutr.,* 29(1): 100-103.

Benedetto-M-di (2003). The effect of organic acids (Acid Lac® Reg. Micro PELLETS) on breeder and on gut health maintenance. *Zootecnica – International.*, 6: 40 –45.

Berchieri, A. Jr, Barrow PA (1996). Reduction in incidence of experimental fowl typhoid by incorporation of a commercial formic acid preparation (Bio-Add™) into poultry feed. *Poultry Science.*, 75: 339-341.

Danicke S, Jeroch H, Bottcher W, Bedford MR, Simon O (1999). Effects of dietary fat type, pentosan level and xylanases on digestability of fatty acids, liver lipids, and vitamin E in broilers. *European Journal of Lipid Science and Technology.*, 101: 90-100.

Danicke S, Halle I, Franke A, Jeroch H (2001). Effect of energy source and xylanase addition on energy metabolism, performance, chemical body composition and total body electrical conductivity (TOBECO) of broilers. *Animal Physiology and Animal Nutrition.*, 85(9-10): 301-313.

Dawson KA, Pirulescu M (1999). Proceedings of the Altech's Asia Pacific Lecture Tour., pp. 75-83.

Eidelsburger U, Kirchgessner M (1994). Effect of organic acid salts in the feed on fattening performance of broilers. *Archiv-fur-Geflugelkundes.*, 58(6): 268 –277.

Fairchild AS, Grimes JL, Jones FT, Wineland MJ, Edens FW, Sefton AE (2001). Effects of hen age, Bio-MOS® and Flavomycin, on Poultry susceptibility to oral *E.coli* challenge. *Poultry Science.*, 80: 562 –571.

Ganguly S, Paul I, Mukhopadhayay SK (2009). Immunostimulants- Their Significance in Finfish Culture. *Fishing Chimes.*, 29(7): 49-50.

Ganguly S, Paul I, Mukhopadhayay SK (2010). Immunostimulant, probiotic and prebiotic – their applications and effectiveness in aquaculture: A Review. *Israeli J. Aquacult. – Bamidgeh.*, 62(3): 130-138

Ganguly S, Paul I, Mukhopadhayay SK (2010). Immunomodulatory Effects of Fungal Äeta - Glucans In Fish Farming. Fishing Chimes., 30(7): 64.

Ganguly S, Prasad A (2012). Microflora in fish digestive tract plays significant role in digestion and metabolism: a *Review. Rev. Fish Biol. Fisheries.*, 22: 11-16, doi: 10.1007/s11160-011-9214-x.

Gatesoupe FJ (2005). Probiotics and prebiotics for fish culture, at the parting of the ways. Aqua Feeds: Formulation & Beyond., 2(3): 3-5

Gatlin DM III, Li P (2004) .Dietary supplementation of prebotics for health management of hybrid striped bass *Morone chrysops* x *M. saxatilis. Aqua Feeds: Formulation & Beyond.*, 1(4): 19-21.

Gibson GR, Roberfroid MB (1995). Dietary modulation of human colonic microbiota, introducing the concept of prebiotics. *Journal of Nutrition.*, 125: 1401-1412.

Hinton M, Linton AH, Perry FG (1985). Control of Salmonella by acid disinfection of chicks feed. *Veterinary Record.*, 116: 502.

Iji PA, Tivey DR (1998). Natural and Synthetic oligosachharides in brolier chicken diets. *World's Poultry Science.*, 54: 129-143.

Izat AL, Adams MH, Cabel MC, Colberg M, Reiber MA, Skinner JT, Waldroup PW, (1990). Effects of formic acid or calcium formate in feed on performance and microbiological characteristics of broilers. *Poultry Science.*, 69: 1876-1882.

Kaniawati S, Skinner JT, Waldroup PW, Izat AL, Colberg M. (1992) Effects of feeding organic acids to broilers on performance and salmonella Colonization of the caeca and or contamination of the carcass. *Poultry Science.*, 71 (Suppl.1): 159.

Li P., Gatlin DM III (2004). Evaluation of the prebiotic Grobiotic®-A and brewers' yeast as dietary supplements for sub-adult hybrid striped bass (*Morone chrysops* × *M. saxatilis*) challenged in situ with *Mycobacterium marinum*. *Aquaculture.*, 248: 197-205.

Loddi MM (2003). Probioticos and prebioticos e acidificanate organico. em dietas para frangos de corte [tese]., Taboticabal. FCAV, UNESP.

Loddi MM, Moraes VM, Nakaghi LSO., Tucci FM, Hannas MI, Ariki JA (2004) . Mannanoligosaccharide and organic acids on performance and intestinal morphometric characteristics of broiler chickens. In: Proceeding of the 20th Annual symposium., Suppl.1: 45.

Lon R (1995). Dietary MOS as an approach for altering prevalence of antibiotic resistance and distribution of tetracycline resistant determinants: in fecal bacteria from swine. MS Thesis, University of Kentucky.

Mairoka A, Santin AME, Borges SA, Opalinski M, Silva AVF (2004). Evaluation of a mix of fumaric, lactic, citric and ascorbic acids on starter diets of broiler. *Archives Veterinary Sciences.*, 9(1): 31 – 37.

Mazurkiewicz J, Przyby³ A, Golski J (2008). Usability of Fermacto prebiotic in feeds for common carp (*Cyprinus carpio* L.) fry. *Nauka Przyr. Technol.*, 2, 3, #15.

Newman K (1994). Mannanoligosaccharides. Natural Polymers with significant impact on the G.I. microflora and the immune system. In: Biotechnology in the feed industry proceeding of Alltech's 10th Annual Symposium., Nottingham University Press, Nottingham, UK. pp. 167 - 174.

Nursey I (1997). Control of Salmonella. Kraftfutter., 10: 415-422.

Parks CW, Grimes JL, Ferket PR, Fairchild AS (2001). The effect of mannanoligosaccharide, bambermycin and virginiamycin on performance of large white male turkey toms. *Poultry Science.*, 80: 718-723.

Pelicano ERL, Souza PA, Souza HBA, Figueiredo DF, Boiago MM, Carvalho SR, Bordon VF (2005). Revista Brasilaria de Cieneia Avicola., 7(4): campinas Oct/Dec.

Radecki SY, Yokoyama MT (1991). Intestinal bacteria and their influence on swine nutrition. In: Miller ER, Duane EU, Lewis AJ. Swine Nutrition. Boston: Butterworth – Heinemann., pp.439-447.

Ringø E, Olsen RE, Gifstad TO, Dalmo RA, Amlund H, Hemre GI, Bakke AM (2010). Prebiotics in aquaculture- a review, *Aquacult. Nutr.*, 16(2): 117-136.

Santin E, Mairoka A, Macari M, Grecco M, Sanchez JC, Okada TM, Myasaka AM (2001). Performance and intestinal mucosa development of broiler chickens fed

Ganguly, S. 2013. Chemical aspects of fermentation technology in food processing industries. *Res. J. Chem. Environ. Sci.* 1(1): 42-3.

Harden, A. and Young, W. J. 1906. The Alcoholic Ferment of Yeast-Juice. *Proceedings of the Royal Society of London.* 78 (526): 369-75.

Ruddle Kenneth, and Naomichi Ishige. 2010. On the Origins, Diffusion and Cultural Context of Fermented Fish Products in Southeast Asia. In: *Globalization, Food and Social Identities in the Asia Pacific Region*, ed. James Farrer. Tokyo: Sophia University Institute of Comparative Culture http://icc.fla.sophia.ac.jp/global%20food%20papers/html/ruddle_ishige.html

Macfarlane G. T. and Macfarlane S. 1993. Factors affecting fermentation reactions in the large bowel. *Proceedings of the Nutrition Society.* 52: 313-61.

Steinkraus, K.H. 1995. Ed. *Handbook of Indigenous Fermented Foods.* NY, Marcel Dekker, Inc.

Yan Lin, Wei Zhang, Chunjie Li, Kei Sakakibara, Shuzo Tanaka and Hainan Kong (2012) Factors affecting ethanol fermentation using *Saccharomyces cerevisiae* BY4742, *Biomass and Bioenergy.* 47: 395-401.

Evenson, M.L., Hinds, M.W., Bernstein, R.S. and Bergdoll, M.S. 1988. Estimation of human dose of Staphylococcal enterotoxin A from a large outbreak of staphylococcal food poisoning involving chocolate milk. *International Journal of Food Microbiology.* 7:311–6.

Govindan, T.K. 1985. Handling, preservation and transportation of fresh fish. pp. 44-76. *In:* Fish Processing Technology. Oxford and IBH Publishing Co., Oxford.

International Commission on Microbiological Specification for Food (ICMSF). 2nd ed. Microbiological analysis: principles and specific applications. University of Toronto press, New York, USA, 1986.

Subburaj M., Karunasagar Indrani and Karunasagar I. 1984. Incidence of histidine decarboxylating bacteria in fish and market environs. *Food Microbiology.* 1: 263-7.

Huss, H.H. 1994. Assurance of seafood quality, FAO, Fisheries technical paper, 334: pp.59, Rome.

Jeyasekaran, G. and Ayyappan, S. 2003. Microbiological quality of farm-reared freshwater fish, rohu (*Labeo rohita*). *Indian Journal of Fish.* 50(4): 455-9.

Karungi, C. Byaruhanga, Y. B. and Muyonga, J. H. 2004. Effect of preicing duration on quality deterioration of iced Nile perch (*Lates niloticus*). *Food Chemistry.* 85: 13-7.

Lakshmanan, R. Jeya Shakila, R. and Jeyasekaran, G. 2002. Survival of amine forming bacteria during the ice storage of fish and shrimp. *Food Microbiology,* 19: 617-25.

Lilabati, H. and Vishwanath, W. 1999. Biochemical and microbiological quality of *Labeo gonius* stored in ice. *Fishery Technology.* 36(1): 24-7.

Nair, K.K.S. and Nair, R.B. 1988. Bacteriological quality of freshwater fish from Krishnarajendra sagar reservoir. *Fishery Technology.* 25: 79-80.

Nair, R.B. Tharamari, P.K. and Lahiry, N.L. 1974. Studies on chilled storage of fresh water fish. II. Factors affecting quality. *Journal of Food Science and Technology*. **11**(1): 118-22.

Ozyurt, G. Kuley, E. Ozkutuk, S. and Ozogul, F. 2009. Sensory, microbiological and chemical assessment of the freshness of red mullet (*Mullus barbatus*) and goldband goatfish (*Upeneus moluccensis*) during storage in ice. *Food Chemistry*, **114**(2): 505-10.

Panda, S.K. and Nayak, B.B. 2001. Analysis of seafood for *Vibrios* pp. 192-195. *In* Quality management in export of seafood products. Central Institute Fisheries Education publication, Mumbai, India.

Ruiz-Capillas, C. and Moral, A. 2005. Sensory and biochemical aspects of quality of whole bigeye tuna (*Thannus obesus*) during bulk storage in controlled atmospheres. *Food Chemistry*. **89**(3): 347-54.

Russel, A.D. and Fuller, R. 1979. Cold tolerant spoilage and their environment. Academic press, UK. pp. 117.

Shewan, J.M. 1962. The bacteriology of fresh and spoiling fish and some related chemical changes, pp. 167-93. *In:* Hawthorn, J. and Leitch, J. M., *Recent advances in food science*. Butterworths, London.

Shewan, J.M. and Murray, C.K. 1979. The microbial spoilage of fish with special reference to the role of psychrophiles, pp. 36-117. *In:* Russell, A. D. and Fuller, R., *Cold tolerant microbes in spoilage and the environment*. Academic Press, London.

Simon S S. and Sanjeev S. 2007. Prevalence of enterotoxigenic *Staphylococcus aureus* in fishery products and fish processing factory workers. *Food Control* **12**: 1565-8.

Sinha, D.K. Choudhary, S.P. and Narayan, K.G. 1991. Microbiological characteristics of marketed rohu (*Labeo rohita*). *Indian Journal of Fish*. **38**: 69-71.

Stansby, M.E. and Lemon, J.M. 1941. Studies on the handling of fresh mackerel (*Scomber scombrus*).U.S. Fish and Wildlife Service, Research report no. 1, pp. 46.

Subburaj, M., *Karunasagar, Indrani and Karunasagar*, I. 1984. Incidence of histidine decarboxylating bacteria in fish and market environs. *Food Microbiology*. **1**: 263-7.

Jablonski, L.M. and Bohach, G.A. 2001. *Staphylococcus aureus. In*: M.P. Doyle, L. Beuchat and T. Montville, (eds), *Food microbiology: fundamentals & frontiers* (2nd ed.), ASM Press, Washington DC. 2001. pp. 411–33.

Sudhi, K.S. 2002. Fish in Kochi markets highly contaminated: A study. Online edition of India's National Newspaper. Kerala, India.

Vieira, R.H.S.F. and Vieira, G.H.F. 1989. Handling of fish products under refrigeration on fishing boats. pp. 3001-4. *In:* Vieira, R.H.S.F. and Vieira, G.H.F. (eds). Science and Technology of Fisheries Products: Methods of Preservation and Transportation of Fisheries Products.

D. Cavalieri, P.E. McGovern, D.L. Hartl, R. Mortimer, M. Polsinelli. Evidence for *S. cerevisiae* fermentation in ancient wine. *Journal of Molecular*

Evolution. **57,** Suppl 1: S226–32, 2003.

K.H. Steinkraus, Ed.*Handbook of Indigenous Fermented Foods.* NY, Marcel Dekker, Inc., 1995.

Ganguly, S. *Food Microbiology.* LAP LAMBERT Academic Publishing GmbH & Co. KG, Saarbrücken, Germany [**ISBN:** 978-3-8484-8217-7], 2012.

Regulation (EC) No.178/2002 of the European Parliament and of the Council of 28 January 2002 laying down the general principles and requirements of food law, establishing the European Food Safety Authority and laying down procedures in matters of food safety.

A. Harden, W. J. Young. The Alcoholic Ferment of Yeast-Juice. *Proceedings of the Royal Society of London,* 1906,**78**(526): 369–375.

J. Dubos. Louis Pasteur: Free Lance of Science, Gollancz. In:K.L. Manchester. Louis Pasteur (1822–1895)-chance and the prepared mind. *Trends Biotechnol.*1995, **13** (12): 511–515.

Nobel Laureate Biography of Eduard Buchner at http://nobelprize.org, The Nobel Prize in Chemistry, 1929.

R. Laudan. In Praise of Fast Food. UTNE Reader. If we fail to understand how scant and monotonous most traditional diets were, we can misunderstand the ethnic foods we encounter in cookbooks, at restaurants, or on our travels, 2010.

R. Laudan. In Praise of Fast Food. UTNE Reader. Where modern food became available, people grew taller and stronger and lived longer, 2010.

Ahmed, S., Dora, K.C.,Sarkar, S.,Chowdhury, S. and Ganguly, S. Isolation and molecular characterization of *Bacillus* species from *Shidal*- a fermented fish product of Assam. *Indian J. Fisheries.*Accepted, 2012a.

Ahmed, S., Dora, K.C.,Sarkar, S.,Chowdhury, S. and Ganguly, S. Quality analysis of shidal- a traditional fermented fish product of Assam. *Indian J. Fisheries.* Accepted, 2012b.

Why does Alaska have more botulism? Centers for Disease Control and Prevention (U.S. federal agency).

Food and Drug Administration. Guide to minimize microbial food safety hazards of fresh-cut fruits and vegetables (PDF), 2010. http://www2a.cdc.gov/phlp/docs/US%20FDA_CFSAN_Food%20Safet.pdf.

Food and Drug Administration. Fish and fisheries products hazards and controls guidance, 3rd ed. Archived, 2007. http://web.archive.org/web/20070929115907/http://www.cfsan.fda.gov/~comm/haccp4.html.

Food and Drug Administration. Managing Food Safety: A HACCP Principles Guide for Operators of Food Establishments at the Retail Level,2007,http://vm.cfsan.fda.gov/~dms/hret-toc.html.

Food Safety and Inspection Service. FSIS Microbiological Hazard Identification Guide For Meat And Poultry Components of Products Produced By Very Small Plants. 2007, http://www.fsis.usda.gov/Frame/FrameRedirect.asp?main=http://www.fsis.usda.gov/oa/haccp/hidguide.htm.

Food Irradiation-A technique for preserving and improving the safety of food. WHO, Geneva (*cited by*Sekhar R. Katta, D.R. Rao, G.R. Sunki, C.B. Chawan. Effect of gamma irradiation of whole chicken carcasses on bacterial loads and fatty Acids. *Journal of Food Science,* 1991, 56 (2), 371-372.

Naik, J., Raju, C.V. and Manjanaik, B. Potential application of irradiation in relation to fish and fishery products. Aquafind-Aquatic fish database est. 1991. (http:// aquafind.com/journalpapers/journal-listings.php).

C.M. Deeley, M. Gao, R. Hunter, D.A.E. Ehlermann, The development of food irradiation in the Asia Pacific, the Americas and Europe; tutorial presented to the International Meeting on Radiation Processing, Kuala Lumpur, 2006. http:// www.doubleia.org/index.php?sectionid=43&parentid=13&contentid=494.

T. Kume, M. Furuta, S. Todoriki, N. Uenoyama and Y. Kobayashi. Status of food irradiation in the world. *Radiation Physics and Chemistry,* 2009,78 (3), 222-226.

Farkas,J. and Mohacsi-*Farkas,* C. (2011). History and future of food irradiation. *Trends in Food Science & Technology,* 2011, 22, 121-126.

Joint FAO/IAEA Division of Nuclear Techniques in Food and Agriculture, IAEA, International Database on Insect Disinfestation and Sterilization – IDIDAS(http:/ /ididas.iaea.org/).

Dosimetry for Food Irradiation, IAEA, Vienna. Technical Reports Series No. 409, 2002.

K. Mehta. Radiation Processing Dosimetry – *A Practical Manual,* GEX Corporation, Centennial, USA, 2006.

J.F. Diehl.*Safety of irradiated foods,* Marcel Dekker, New York. 2nd ed., 1995, pp 291-293.

World Health Organization. Wholesomeness of irradiated food. Geneva, *Technical Report* Series No. 659, 1981.

World Health Organization. Safety and Nutritional Adequacy of Irradiated Food. Geneva, Switzerland: World Health Organization (*cited by* Katherine M. Shea and the Committee on Environmental Health (2000). *Technical Report: Irradiation of Food. Pediatrics,* 1994, 106 (6), 1505-1510).

US Department of Health, and Human Services, Food, and Drug Administration. Irradiation in the production, processing, and handling of food. *Federal Register,* 1986, 51, 13376-13399.

M.T. Osterholm and M.E. Potter. Irradiation pasteurization of solid foods: Taking food safety to the next level, *Emerging Infectious Diseases,* 1997, 3, 4, 575-577.

J.B. Bender, K.E. Smith, C. Hedberg and M.T. Osterholm. Food-borne disease in the 21st century. What challenges await us? *Postgraduate Medicine,* 1999, 106, 2, 109-112, 115-116, 119.

M.T. Osterholm and A.P. Norgan. The role of irradiation in food safety, *The New England Journal of Medicine,* 2004, 350, 18, 1898–1901.

Ganguly, S. and Mukhopadhayay, S.K. Use of food irradiation in food processing industries. Proceedings of International Conference on "Search for a holistic combination of Agriculture, Industry and Education" organized by Dept of Commerce, St. Xavier's College (Autonomous), Kolkata and Dept of Commerce, Netaji Nagar College at St. Xavier's College Auditorium, Kolkata, India (Dec 8, 2011), 2011, pp.25.

Ganguly, S.,Mukhopadhayay, S.K. and Biswas, S.Preservation of food items by irradiation process. *Inter. J. Chem&Biochem. Sci.*, 2012, 1, 11-13.

Prasad A, Ganguly S Herbal Immunomodulators. AV Akademikerverlag GmbH & Co. KG, Saarbrücken, Germany *with trademark* LAP LAMBERT Academic Publishing. 2012;ISBN 978-3-659-30401-9.

Kolte AY, Siddiqui MF, Mode SG. Immunomodulating effect of *Withania somnifera* and *Tinospora cordifolia* in broiler birds. 2007.

Rege NN, Dahanukar SA. Quantitation of microbicidal activity of mononuclear phagocytes : an in vitro technique. *J. Postgrad Med* 1993;39(1): 22-25.

Rege NN, Nazareth HM, Bapat RD, Dhanukar SA. Modulation of immunosuppression in obstructive jaundice by *Tinospora cordifolia*. *Indian J Med Res* 1989;90: 478-83.

Bishavi B, Roychowdhury S, Ghosh S, Sengupta M. Hepatoprotective and immunomodulatory properties of *Tinospora cordifolia* in CCl_4 in toxicated mature albino rats. *J Toxicol* Sci 2002;27(3): 139-46.

Manjrekar PN, Jolly CI, Narayan S. Comparative studies of the immunomodulatory acivities of *Tinospora cordifolia* and *Tinospora sinensis*. Fitoterapia 1999;71: 254-57.

Krishnamohan AV, Reddy DB, Sarma B, John Kirubharan J. Studies on the effects of levamisole against Newcastle disease virus in chicken. *Indian J Comp Microbiol Immunol Infect Dis*. 1997;8: 1-6.

Kujur RT. Evaluation of certain immunomodulatory agents in countering immunosuppressive effects of vaccine strain of infectious bursal disease virus in chicks. 2001;MVSc thesis. Rajendra Agricultural Univ., Bihar, India.

Kumar P. Studies on comparative immunomodulatory effect of herbal preparation and Vitamin E-Se in comparison to Levamisole in broiler chicks. 2003;MVSc thesis. Birsa Agricultural Univ., Ranchi, India.

Nadkarni AK. Indian Materia Medica. Popular Book Depot, Bombay. 3 rd ed., 1954;1: 153-55.

Kuttan G, Kuttan R. Immunomodulatory activity of a peptide isolated from *Viscum album* extract. *Immunol Invest* 1992;21: 285-96.

Ganguly S. and Prasad A. Role of plant extracts and cow urine distillate as immunomodulator in comparison to levamisole– a Review. *J Immunol Immunopathol* 2010;12(2): 91-94

Aney JS, Tambe R, Kulkarni M, Bhise K. Pharmacological And Pharmaceutical Potential of *Moringa oleifera*: A Review. *Journal of Pharmacy Research* 2009;2(9):1424-26.

Prasad, A, Ganguly S. Promising medicinal role of *Moringa oleifera*: a Review. *J Immunol Immunopathol* 2012;14(1): 1-5. DOI: 10.5958/j.0972-0561.14.1.001.

Michael L. Bioremediation of Turbid Surface Water Using Seed Extract from *Moringa oleifera* Lam. (Drumstick) Tree. 2010;doi:10.1002/9780471729259.mc01g02s16.

Akhouri S, Prasad A, Ganguly S. *Moringa oleifera* leaf extract imposes better feed utilization in broiler chicks. *J Biol Chem Res* 2013;30(2): 447-50.

Ramachandran C, Peter KV, Gopalakrishnan PK. Drumstick (*Moringa oleifera*): A multipurpose Indian Vegetable. *Economic Botany* 1980;34(3): 276-83.

Ganguly, S. Health benefits of coconut in the Asian cuisines: A Review. *J Biol Chem Res* 2013a;30(2): 517-21.

Nneli, RO, Woyike, OA. Anti-ulcerogenic effects of coconut (*Cocos nucifera*) extract in rats. *Phytother Res* 2008;22: 970-72.

Mensink, Ronald P, Peter L Zock, Arnold DM Kester and Martijn B Katan. Effects of dietary fatty acids and carbohydrates on the ratio of serum total to HDL cholesterol and on serum lipids and apolipoproteins: a meta-analysis of 60 controlled trials. *American Journal of Clinical Nutrition* (American Society for Clinical Nutrition). 2003;77(5): 1146-55.

Ganguly, S. Medicinal properties of lime and its traditional food value. Res J Pharm Sci 2013b;2(4): 19-20.

Room Adrian. A dictionary of true etymologies, Taylor & Francis. 1986;pp. 101.

Raichlen Steven. Small citruses yield tart juice, aromatic oils, big, fresh taste. 1992;The Baltimore Sun.

Ganguly S. Herbal and plant derived natural products as growth promoting nutritional supplements for poultry birds: A Review. *J Pharm Sci Innov* 2013c; 2(3): 12-13. DOI: 10.7897/2277-4572.02323.

Ganguly S. A Handbook on Traditional Medicinal Plants, Herbs and Fruits in Indian Agriculture and Forestry. 1st ed. International E-Publication. 2013d;ISBN 978-81-927544-5-1. Official e-Book (Section: Agriculture & Forestry Sciences) publication of the International Science Congress Association, Indore, UP, India.

Banerjee S, Mukhopadhayay SK, Haldar S, Ganguly S, Pradhan S, Patra S, Niyogi D Isore DP. Effect of phytogenic growth promoter on broiler birds. *J Pharmacogn Phytochem* 2013a;1(6): 183-88.

Banerjee S, Mukhopadhayay SK, Ganguly S. Phytogenic growth promoter as replacers for antibiotic growth promoter in poultry birds. *J Anim Genet Res* 2013b;1(1): 6-7, DOI: 10.12966/jagr.05.02.2013.

Bohn J. A. and BeMiller J. N. 1995. (1ÀÛÆÜ3)-â-glucans as biological response modifiers: a review of structure-functional activity relationships. *Carbohydrate polymers*. 28: 3-14.

Brown G. D. and Gordon S. 2003. Fungal beta-glucans and mammalian immunity. *Immunity*. 19: 311-15.

Engstad C. S., Engstad R. E., Olsen J. O. and Osterud B. 2002. The effect of soluble beta-1,3- glucan and lipopolysaccharide on cytokine production and coagulation activation in whole blood. *International Immunopharmacology*. 2: 1585-97.

Ganguly S., Paul I. and Mukhopadhayay S.K. 2009. Immunostimulants- Their Significance in Finfish Culture. *Fishing Chimes*. 29(7): 49-50.

Ganguly S., Paul I. and Mukhopadhayay S.K. 2010. Immunomodulatory Effects of Fungal Âeta - Glucans In Fish Farming. *Fishing Chimes*. 30(7): pp. 64.

Ganguly S., Dora K. C., Sarkar S. and Chowdhury S. 2013a. Supplementation of prebiotics in fish feed- A Review. *Rev. Fish Biol. Fisheries*. 23(2): 195-99, DOI: 10.1007/s11160-012-9291-5.

Ganguly S. 2013b. Fundamentals of Fish Immunostimulants. Research India Publications, Delhi. ISBN 978-81-89476-07-6.

Guo Y., Ali R. A. and Qureshi M. A. 2003. The influence of beta-glucan on immune responses in broiler chicks. *Immunopharmacology and Immunotoxicology*. 25: 461-72.

Huff G. R., Huff W. E., Rath N. C. and Tellez G. 2006. Limited Treatment with â-1,3/1,6-Glucan Improves Production Values of Broiler Chickens Challenged with *Escherichia coli. Poult. Sci.* 85:613–18.

Lowry V. K., Farnell M. B., Ferro P. J., Swaggerty C. L., Bahl A. and Kogut M. H. 2005. Purified beta-glucan as an abiotic feed additive up-regulates the innate immune response in immature chickens against *Salmonella enterica* serovar *Enteritidis*. *International Journal of Food Microbiology*. 98: 309-18.

Onarheim A. M. 1992. Now a yeast extract to fortify fish. *Fish Farmer*. 15: pp. 45.

Paul I, Isore D. P., Joardar S. N., Samanta I., Biswas U., Maiti T. K., Ganguly S. and Mukhopadhayay S. K. 2012. Orally administered α-glucan of edible mushroom (*Pleuratus florida*) origin upregulates immune response in broiler. *Indian J. Anim. Sci.* 82(7): 745-48

Paul I., Isore D. P., Joardar S. N., Roy B., Aich R. and Ganguly S. 2013. Effect of dietary yeast cell wall preparation on innate immune response in broiler chickens. *Indian J. Anim. Sci.* 83(3): 307-09.

Persson Waller K., Gronlund U. and Johannisson A. 2003. Intramammary infusion of beta1,3- glucan for prevention and treatment of *Staphylococcus aureus* mastitis. *J. Vet. Med.* B. Infect. Dis. Vet. Public Health. 50: 121-27.

Reynolds J. A., Kastello M. D., Harrington D. G., Crabbs C. L., Peters C. J. and Jemski J. V. 1980. Glucan-induced enhancement of host resistance to selected infectious diseases. *Infection and Immunity*. 30: 51-57.

Rice P. J., Adams E. L., Ozment-Skelton T., Gonzalez A. J., Goldman M. P. and Lockhart B. E. 2005. Oral delivery and gastrointestinal absorption of soluble glucans stimulate increased resistance to infectious challenge. *The journal of pharmacology and experimental therapeutics*. 314: 1079-86.

Seljelid R. 1990. Immunomodulators-medicine for the 90is In: Pathogenesis of wound and biomaterial associated infections (ed. Wadstrom T, Eliasson I, Holder I and Ljungh A.). Springer-verlag, Berlin. pp. 107-13.

Selvaraj V., Sampath K. and Sekar V. 2005. Administration of yeast glucan enhances survival and some non-specific and specific immune parameters in carp (*Cyprinus carpio*) infected with *Aeromonas hydrophila*. *Fish Shellfish Immunol*. 19: 293–306.

Tsukada C., Yokoyama H., Miyaji C., Ishimoto Y., Kawamura H. and Abo T. 2003. Immunopotentiation of intraepithelial lymphocytes in the intestine by oral administrations of beta-glucan. *Cellular Immunology*. 221: 01-05.

Vetvicka V., Terayama K., Mandeville R., Brousseau P., Kournikakis B. and Ostroff G. 2002. Orally-administered yeast beta-1,3-glucan prophylactically protects against anthrax infection and cancer in mice. *The Journal of the American Nutraceutical Association*. 5(2): 1- 5.

Brown G. D. and Gordon S. 2003. Fungal beta-glucans and mammalian immunity. *Immunity*. 19: 311-15.

Wakshull E., Brunke-Reese D., Lindermuth J., Fisette L., Nathans R. S. and Crowley J. J. 1999. PGG-glucan, a soluble beta-(1,3)-glucan, enhances the oxidative burst response, microbicidal activity, and activates an NF-êB-like factor in human PMN: evidence for a glycosphingolipid beta-(1,3)-glucan receptor. *Immunopharmacology*. 41: 89-107.

Waller K. P. and Colditz I. G. 1999. Effect of intramammary infusion of beta-1,3-glucan or interleukin-2 on leukocyte subpopulations in mammary glands of sheep. *Amer. J. Vet. Res*. 60: 703-07.

Williams D. L. and Diluzio N. R. 1979. Glucan induced modification of experimental Staphylococcus aureus infection in normal, leukemic and immunosuppressed mice. *Adv. Exp. Med. Biol*. 121: 291-306.

Xiao Z., Trincado C. A. and Murtaugh M. P. 2004. Beta-glucan enhancement of T cell INFgamma response in swine. Vet. Immunol. *Immunopathol*. 102: 315-20.

Yun C. H., Estrada A., Van Kessel A, Park B. C. and Laarveld B. 2003. Beta-glucan, extracted from oat, enhances disease resistance against bacterial and parasitic infections. *FEMS Immunology and Medical Microbiology*, 35: 67-75.

Seljelid R. Immunomodulators-medicine for the 90is In: Pathogenesis of wound and biomaterial associated infections (ed. Wadstrom T, Eliasson I, Holder I and Ljungh A.). Springer-verlag, Berlin. 1990;107-13.

Onarheim A M. Now a yeast extract to fortify fish. *Fish Farmer*. 1992;15: 45.

Bohn J A, BeMiller J N. (1ÀÛÆÜ3)-α-glucans as biological response modifiers: a review of structure-functional activity relationships. *Carbohydrate polymers*. 1995;28: 3-14.

Guo Y, Ali R A, Qureshi M A, The influence of beta-glucan on immune responses in broiler chicks. *Immunopharmacology and Immunotoxicology*. 2003;25: 461-72.

Lowry V K, Farnell M B, Ferro P J, Swaggerty C L, Bahl A, Kogut M H. Purified beta-glucan as an abiotic feed additive up-regulates the innate immune response in immature chickens against Salmonella enterica serovar Enteritidis. *International Journal of Food Microbiology*. 2005;98: 309-18.

Brown G D, Gordon S. Fungal beta-glucans and mammalian immunity. *Immunity*.2003;19: 311-315.

Selvaraj V, Sampath K, Sekar V. Administration of yeast glucan enhances survival and some non-specific and specific immune parameters in carp (*Cyprinus carpio*) infected with *Aeromonas hydrophila*. *Fish Shellfish Immunol*. 2005;19: 293–306.

Williams D L, Diluzio N R. Glucan induced modification of experimental Staphylococcus aureus infection in normal, leukemic and immunosuppressed mice. *Adv. Exp. Med. Biol*. 1979;121: 291-306.

Reynolds J A, Kastello M D, Harrington D G, Crabbs C L, Peters C J, Jemski J V. Glucan-induced enhancement of host resistance to selected infectious diseases. *Infection and Immunity*. 1980;30: 51-57.

Xiao Z, Trincado C A, Murtaugh M P. Beta-glucan enhancement of T cell INFgamma response in swine. Vet. Immunol. *Immunopathol*. 2004;102: 315-20.

Banga R K, Singhal L K, Chauhan R S Cow urine and immunomodulation: An update on cowpathy. *Int. J. Cow Sci*. 2005;1(2): 26-29.

Barrow P A, Wallis T S Vaccination against Salmonella infections in food animals: Rationale, Theoretical Basis and Practical application. Eds. Wray, Wray A. Salmonella in Domestic Animals @ CAB International, 2000.

Chauhan R S, Singh B P, Singh G K Immunomodulation with Kamdhenu Ark in mice. *J. Immunol. & Immunopath*. 2001;71: 89-92.

Kumar P, Singh G K, Chauhan R S, Singh D D. Effect of cow urine on lymphocyte proliferation in developing stages of chicks. *The Indian cow*. 2004;2: 3-5.

Garg N, Chauhan R S, Kumar A Assessing the effect of cow urine on immunity of White Leghorn layers. XIIth-ISAH-Congress-on-Animal-Hygiene, Warsaw, Poland, 4 – 8 Sept. 2005. 2: 81-83.

Ganguly S, Paul I, Mukhopadhayay SK. Immunostimulants- Their Significance in Finfish Culture. *Fishing Chimes*. 2009;29(7): 49-50.

Ganguly S, Paul I, Mukhopadhayay SK. Immunomodulatory Effects of Fungal Âeta - Glucans In Fish Farming. *Fishing Chimes* . 2010;30(7): pp. 64.

Ganguly S, Dora K C, Sarkar S, Chowdhury S Supplementation of prebiotics in fish feed- A Review. *Rev. Fish Biol. Fisheries*. 2013a;23(2): 195-99, DOI: 10.1007/s11160-012-9291-5.

Ganguly S. Fundamentals of Fish Immunostimulants. Research India Publications, Delhi [RIP]. 2013b;ISBN 978-81-89476-07-6.

Persson Waller K, Gronlund U, Johannisson A. Intramammary infusion of beta1,3-glucan for prevention and treatment of Staphylococcus aureus mastitis. *J. Vet*.

Med. B. Infect. Dis. Vet. Public Health. 2003;50: 121-127.

Engstad C S, Engstad R E, Olsen J O, Osterud B. The effect of soluble beta-1,3- glucan and lipopolysaccharide on cytokine production and coagulation activation in whole blood. *International Immunopharmacology*. 2002;2: 1585-1597.

Yun C H, Estrada A, Van Kessel A, Park B C, Laarveld B. Beta-glucan, extracted from oat, enhances disease resistance against bacterial and parasitic infections. FEMS *Immunology and Medical Microbiology*, 2003;35: 67-75.

Huff G R, Huff W E, Rath N C. and Tellez G. Limited Treatment with α-1,3/1,6- Glucan Improves Production Values of Broiler Chickens Challenged with *Escherichia coli*. *Poult. Sci*. 2006;85:613–18.

Rice P J, Adams E L, Ozment-Skelton T., Gonzalez A J, Goldman M P, Lockhart B E. Oral delivery and gastrointestinal absorption of soluble glucans stimulate increased resistance to infectious challenge. *The journal of pharmacology and experimental therapeutics*. 2005;314: 1079-86.

Vetvicka V, Terayama K, Mandeville R, Brousseau P, Kournikakis B, Ostroff G. Orally-administered yeast beta-1,3-glucan prophylactically protects against anthrax infection and cancer in mice. *The Journal of the American Nutraceutical Association*. 2002; 5(2): 1- 5.

Ganguly S, Prasad A. Role of plant extracts and cow urine distillate as immunomodulator in comparison to levamisole – a Review. *J. Immunol. Immunopathol* 2010,12(2): 91-94.

Ganguly S, Prasad A. Role of plant extracts and cow urine distillate as immunomodulators: a review. *J. Medi. Pl. Res*. 2011;5(4): 649-51.

Ambwani S, Ambwani T, Singhal L, Chauhan R S. Immunomodulatory effect of cow urine on Dimethoate induced immunotoxicity in Avian Lymphocyte. *The Indian Cow*. 2005;3: 49-51.

Garg N, Kumar A, Chauhan R S Effect of indigenous Cow urine on nutrient utilization of White Leghorn layers. *Int. J. Cow Sci*. 2005;1: 36-38.

Kumar P, Singh G K, Chauhan R S, Singh D D. Effect of cow urine on lymphocyte proliferation in developing stages of chicks. *The Indian cow*. 2004;2: 3-5.

Jojo, R, Prasad A, Tiwary B K, Ganguly, S. Role of cow urine distillate as a potential immunomodulator in broilers: A Research report. *Poult. Line*, 2011;11(12): pp. 27.

Awadhiya R P, Vegad J L, Kalte G N. Dinitrochlorobenzen skin hypersensitivity reaction in the chicken; 1981.

Srikumar R, Parthsarthy N J, Mankandan S, Narayanan G S, Devi R S. *Molecular and Cellular Biochemistry*. 2006;283: pp. 67.

Kumari R. Effect of Asparagus racemosus feeding on humoral and cell mediated immune response in broiler chicks. 2007;MVSc Thesis, Birsa Agricultural University, Ranchi.

Rakhi. Studies on effect of feeding *Withania somnifera* (Ashwagandha) on humoral and cell mediated immune response in broiler chicks. 2004;M.V.Sc. Thesis, Birsa

Agricultural University, Ranchi

Chauhan R S, Singh D D, Singhal L K, Kumar R. Effect of cow urine on IL-1 & IL-2. *J. Immunol. Immunopath*. 2004;6: 38-39.

Brown G D, Gordon S. Fungal beta-glucans and mammalian immunity. *Immunity*. 2003;19: 311-315.

Paul I, Isore D P, Joardar S N, Samanta I, Biswas U, Maiti T K, Ganguly S, Mukhopadhayay S K. Orally administered â-glucan of edible mushroom (*Pleuratus florida*) origin upregulates immune response in broiler. Indian J. Anim. Sci. 2012; 82(7): 745-48

Waller K P, Colditz I G. Effect of intramammary infusion of beta-1,3-glucan or interleukin-2 on leukocyte subpopulations in mammary glands of sheep. Amer. *J. Vet. Res*. 1999;60: 703-07.

Tsukada C, Yokoyama H, Miyaji C, Ishimoto Y, Kawamura H, Abo T. Immunopotentiation of intraepithelial lymphocytes in the intestine by oral administrations of beta-glucan. *Cellular Immunology*. 2003;221: 01-05.

Burt S. Essential oils: their antibacterial properties and potential applications in foods—a review. *Int. J. Food Microbiol* 2004,94(3): 223-53.

Aksit M, Goksoy E, Kok F, Ozdemir D, Ozdogan M. The impacts of organic acid and essential oil supplementations to diets on the microbiological quality of chicken carcasses. Arch. Geflugelkd 2006,70: 168–73.

Batal A B, Parsons C M. Effect of age on nutrient digestibility in chicks feed different diets. *Poultry Science* 2002,81: 400-407.

Jamroz D, Wertelecki T, Houszka M, Kamel C. Influence of diet type on the inclusion of plant origin active substances on morphological and histochemical characteristics of the stomach and jejunum walls in chicken. *J. Anim. Physiol. Anim. Nutr*. (Berl) 2006,90: 255–68.

Jamroz D, Kamel C. Plant extracts enhance broiler performance. *J. Anim. Sci* 2002,80: 41.

Biavatti M W et al. Preliminary studies of alternative feed additive for broilers: Alternanthera brasiliana extract, propolis extract and linseed oil. *Rev. Brasscienc. Avic* 2003,5: 147-51.

Hernandez F, Madrid J, Garcia V, Orengo J, Megias M D. Influence of two plant extracts on broilers performance, digestibility, and digestive organ size. *Poultry Science* 2004,83: 169-74.

Jang I S et al. Effect of a commercial essential oil on growth performance, digestive enzyme activity and intestinal microflora population in broiler chicks. *Animal Feed Sci. and Tech* 2006,134: 305-15.

Banerjee S, Mukhopadhayay SK, Haldar S, Ganguly S, Pradhan S, Patra S, Niyogi D, Isore DP. Effect of phytogenic growth promoter on broiler birds. *Indian J. Vet. Pathol*. 2013,37(1): 34-37.

Ganguly S. Cow urine distillate is regarded as promising immunomodulatory. supplement for broiler diet: A Review. *Unique J Ayurvedic Herbal Medi.* 2013;1(1): 03-04.

Khanuja S P S. Use of bioactive fraction from cow urine distillate ('Go-mutra') as a bio-enhancer of anti-infective, anti-cancer agents and nutrients, 2007 (www.freepatentsonline.com/7235262.html).

Ganguly S, Prasad A. Role of plant extracts and cow urine distillate as immunomodulator in comparison to levamisole – a Review. *J. Immunol. Immunopathol* 2010;12(2): 91-4.

Burt S. Essential oils: their antibacterial properties and potential applications in foods—a review. *Int. J. Food Microbiol* 2004;94(3): 223-53.

Aksit M, Goksoy E, Kok F, Ozdemir D, Ozdogan M. The impacts of organic acid and essential oil supplementations to diets on the microbiological quality of chicken carcasses. Arch. Geflugelkd 2006,70: 168–73.

Batal A B, Parsons C M. Effect of age on nutrient digestibility in chicks feed different diets. *Poultry Science* 2002;81: 400-7.

Jamroz D, Wertelecki T, Houszka M, Kamel C. Influence of diet type on the inclusion of plant origin active substances on morphological and histochemical characteristics of the stomach and jejunum walls in chicken. *J. Anim. Physiol. Anim. Nutr.* (Berl) 2006;90: 255–68.

Jamroz D, Kamel C. Plant extracts enhance broiler performance. J. Anim. Sci 2002;80: 41.

Biavatti M W *et al.* Preliminary studies of alternative feed additive for broilers: Alternanthera brasiliana extract, propolis extract and linseed oil. Rev. Brasscienc. Avic 2003;5: 147-51.

Hernandez F, Madrid J, Garcia V, Orengo J, Megias M D. Influence of two plant extracts on broilers performance, digestibility, and digestive organ size. *Poultry Science* 2004;83: 169-74.

Jang I S et al. Effect of a commercial essential oil on growth performance, digestive enzyme activity and intestinal microflora population in broiler chicks. *Animal Feed Sci. and Tech* 2006;134: 305-15.

Adebiyi FM, Akpan I, Obiajunwa EI, Olaniyi HB. Chemical/Physical Characterization of Nigerian Honey. *Pakistan Journal of Nutrition.* **2004**;3(5): 278-81.

Rodriguez OG. Characterization of honey produced in Venezuela. *Food Chemistry.* **2004**;84(4): 499-502.

Rehman S, Khan FZ, Maqbool T. Physical and spectroscopic characterization of Pakistani honey. *Chem Inv. Agr.* **2008**;35(2): 199-204.

Bogdanov S, Jurendic TJ, Sieber R, Gallmann P. Honey for Nutrition and Health: a Review. After: American *Journal of the College of Nutrition.* **2008**;27: 677-89.

Venugopal S, Devaranjan S. Estimation of total flavonoid, phenols and antioxidant activity of local and New Zealand. *Journal of Pharmacy Research.* **2011**;4(2), 464-6.

Alisi CS, Ojiako OA, Igwe CU, Ujowundu CO, Anugweje K, Okwu GN. Antioxidant Content and Free Radical Scavenging Activity of Honeys of Apis mellifera of Obudu Cattle Ranch. *International Journal of Biochemistry Research & Review*. **2012**;2(4): 164-75.

Agbagwa OE, Otokunefor TV, Nnenna FP. Quality assessment of Nigeria honey and manuka honey. *J. Microbiol. Biotech. Res.* **2011**;1(3), 20-31.

Gheldof N, Wang XH, Engeseth NJ. Identification and quantification of antioxidant components of honeys from various floral sources. *Journal of Agricultural and Food Chemistry*. **2002**;50(21): 5870-7.

Makawi SZA, Gadkariem EA, Ayoub SMH. Determination of Antioxidant Flavonnoids in Sudanese Honey Samples by Solid Phase Extraction and High Performance Liquid Chromatography. *E-Journal of Chemistry*, **2009**;6(S1): S429-37.

Candiracci M, Piatti E, Dominguez-Barragan M, Garcia-Antras D, Morgado B, Ruano D, Gutierrez JF, Parrado J, Castano A. Anti-inflammatory Activity of a Honey Flavonoid Extract on Lipopolysaccharide-Activated N13 Microbial Cells. *Journal of Agricultural and Food Chemistry*. **2012**;60: 12304-11.

Nagai T, Inoue R, Kanamori N. Characterization of honey from different floral sources. Its functional properties and effects of honey species on storage of meat. *Food Chemistry*. **2006**; pp.256-62.

Nanda V, Sarkar BC, Sharma HK, Bawa AS. Physico-chemical properties and estimation of mineral content in honey produced from different plants in Northern India. *Journal of Food Composition and Analysis*. **2003**;16: 613-9.

Terrab A, Gonzale MMI, Gonzalez AG. Characterization of Moroccan Unifloral honeys using multivariate analysis. *Euro Food Res Technol*. **2003**;218: 88-95.

Gheldof N, Engeseth NJ. Antioxidant capacity of honeys from various floral sources based on the determination of oxygen radical absorbance capacity and inhibition of invitro lipoprotein oxidation. *J. Agric Food Chem*. **2002**;50: 3050-5.

Mondragon-Cortez P, Ulloa JA, Rosas-Ulloa P, Rodriguez-Rodriguez R, Resendiz Vazquez JA. Physicochemical characterization of honey from the West region of Mexico. *Journal of Food*. **2012**;11: pp.1, 7-13.

Markov K, Major N, Krpan M, Ursulin-Trastenjak N, Hruskar M, Vahcic N. Changes of Antioxidant Activity and Phenolic Content in Acacia and Multifloral Honey During Storage. *Food Technol. Biotechnol*. **2012**;50(4): 434-41.

Bertoncelj J, Dobersek U, Jamnik M, Golob T. Evaluation of the total phenolic content, antioxidant activity and color of Slovenian honey. *Food Chemistry*. **2007**;105: 822-8.

Atoui AK, Mansouri A, Boskou G, Kefelas P. Tea and herbal infusions: their antioxidant activity and phenolic profile. *Food Chemistry*. **2005**;89: 27-36.

Cao GHM, Cutler RG. Oxygen-radical absorbance capacity assay for antioxidants. *Free Radical Biol*. Med. **1993**;14, 303-11.

Aryouet-Grand A, Vennat B, Pouratt A, Legret P. Standardisation dun extra it de propolis at identification des proncipaux constituents. *Journal de pharmacie de Belgique.* **1994**;49: 462-8.

Velaquez E, Tourniar HA, Saavedra G, Schinella GR. Antioxidant activity of Paraguayan plant extract. *Fitoterapia.* **2003**;74: 91-7.

Meda A, Charles EL, Romito M, Millogo J. Determination of the total phenolic, flavonoid and proline content in Burkina Fasan honey, as well as their radical scavenging activity. *Food chemistry.* **2005**;91: 571-7.

Chang C, Yang MH, Wen HM, Chern JC. Estimation of total flavonoid content in propilis by two complementary colometric methods. *Journal of Food and Drug Analysis.* **2002**;10: 178-82.

Bertoncelj J, Dobersek U, Jamnik M, Golob T. Evaluation of the total phenolic content, antioxidant activity and color of Slovenian honey. *Food Chemistry,* **2007**;105: 822-8.

Atoui AK, Mansouri A, Boskou G, Kefelas P. Tea and herbal infusions: their antioxidant activity and phenolic profile. *Food Chemistry.* **2005**;89: 27-36.

Aljadi A, Kamaruddin M. Evaluation of the phenolic contents and antioxidants capacities of two Malaysian floral honeys. *Food Chemistry.* 2004;85(4): 513-8.

Allen KL, Molan PC, Reid GM. The variability of the antibacterial activity of honey. *Apiacta.* **1991**;26(4): 114-21.

Alvarez-Suarez JM, Tulipani SR-E., Battino M. Contribution of honey in nutrition and human health: a review. *Mediterranean Journal of Nutrition and Metabolism.* **2010**;3: 15-23.

Brooks FP. The pathophysiology of peptic ulcer disease. D*igestive Disease and sciences,* **1995**;30(11): 15S-29S.

Cheeseman KH, Slater TF. An introduction to free radical biochemistry. *British Medical Bulletin.* **1991**;49: 481-93.

White JW, Graham JM. Honey. In: The Hive and the Honey Bee, **1992**;869-927.

Estevinho L, Pereira AP, Moreira L, Dias LG, Pereira E. Antioxidant and Intimicrobial of phenolic compound extract of Northeast Portugal honey. *Food and Chemical Toxicology.* **2008**;46: 3774-9.

Molan PC. Authenticity of honey. In P.R. Ashurst & M. J. Dennis (Eds.), Food Authentication, London: Blackie Academic and Professional. **1996**; pp. 259-303.

Molan PC. Re-introducing honey in the management of wounds and ulcers – theory and practice. Ostomy/Wound Management. **2002**;48(11): 28-40.

Auclair C, Viosin E. Red Blood Cells from Patient with Malignant Superoxide Dismutase, Catalase, and Glutathione Peroxidase in Renee Gonzales. *Cancer Res.* **1984**;44: 4137-9.

Bar N, Mukhopadhayay SK, Ganguly S, Pradhan S, Patra NC, Pal S, Goswami J, Singh YD, Halder S. Study on probiotic effect of xylanase supplementation in broiler feed. *Indian Journal of Animal Nutrition* 2012; 29(1): 100-103.

Benedetto-M-di., 2003. The effect of organic acids (Acid Lac® Reg. Micro PELLETS) on breeder and on gut health maintenance. *Zootecnica – International* 2003; 6: 40 –45.

Berchieri A Jr Barrow PA. Reduction in incidence of experimental fowl typhoid by incorporation of a commercial formic acid preparation (Bio-Add™) into poultry feed. **Poultry Science** 1996; 75: 339-41.

Danicke S, Jeroch H, Bottcher W, Bedford MR, Simon O. 1999. Effects of dietary fat type, pentosan level and xylanases on digestability of fatty acids, liver lipids, and vitamin E in broilers. European Journal of Lipid Science and Technology 1999; 101: 90-100.

Danicke S, Halle I, Franke A, Jeroch H. 2001. Effect of energy source and xylanase addition on energy metabolism, performance, chemical body composition and total body electrical conductivity (TOBECO) of broilers. Animal Physiology & Animal Nutrition 2001; 85(9-10): 301-13.

Das D, Mukhopadhyayay SK, Ganguly S, Kar I, Dhanalakshmi S, Singh YD, Singh KS, Ramesh S, Pal S. Mannan oligosaccharide and organic acid salts as dietary supplements for Japanese quail (*Coturnix Coturnix Japonica*). *Int. J Livest Res* 2012; 2(3): 211-14.

Dawson KA, Pirulescu M. Proceedings of the Altech's Asia Pacific Lecture Tour 1999; pp.75-83.

Eidelsburger U, Kirchgessner M. 1994: Effect of organic acid salts in the feed on fattening performance of broilers. Archiv-fur-Geflugelkundes 1994; 58(6): 268-77.

Fairchild AS, Grimes JL, Jones FT, Wineland MJ, Edens FW, Sefton AE. 2001. Effects of hen age, Bio-MOS® and Flavomycin on Poult susceptibility to oral *E. coli* challenge. Poultry Science 2001; 80: 562 –71.

Ganguly S. Supplementation of prebiotics in poultry feed: A Review. *World's Poultry Science Journal*. 2013a; Vol. 69: xx-xx. doi: 10.1017/1 S0043933913000640. In press.

Ganguly S. Herbal and plant derived natural products as growth promoting nutritional supplements for poultry birds: A Review. *J Pharm Sci Innov* 2013b; 2(3): 12-13. DOI 10.7897/2277-4572.02323.

Ganguly, S. A Handbook on Experimentally Proven Non-Antibiotic Growth Promoters as Feed Additives in Animal Nutrition. 1st ed. International E-Publication 2013c;ISBN 978-81-927544-3-7. An official E-Book (Section: Animal, Veterinary & Fishery Sciences) publication of the International Science Congress Association, Indore, UP, India.

Ganguly S. Dora KC, Sarkar S, Chowdhury S. Supplementation of prebiotics in fish feed- A Review. *Reviews in Fish Biology and Fisheries* 2013; 23(2): 195-199, DOI: 10.1007/s11160-012-9291-5.

Ganguly S. Mukhopadhayay SK. Immunostimulants, Probiotics and Prebiotics. LAP LAMBERT Academic Publishing GmbH & Co. KG, Saarbrücken, Germany, 2011; ISBN 978-3-8454-0271-0.

Ganguly S, Paul I, Mukhopadhayay SK. Immunostimulant, probiotic and prebiotic – their applications and effectiveness in aquaculture: A Review. *Israeli J. Aquacult. – Bamidgeh* 2010; 62(3): 130-38.

Gao F, Jiang Y, Zhou GH, Han ZK. 2008. The effect of xylanase supplementation on performance characteristics of G.I. tract, blood parameters and gut microflora in broiler feed on wheat based diet. *Animal Feed Science and Technology* 2008; 142: 173-84.

Gibson GR, Roberfroid MB. Dietary modulation of human colonic microbiota, introducing the concept of prebiotics. *Journal of Nutrition* 1995; 125: 1401-12.

Hinton M, Linton AH, Perry FG. 1985. Control of Salmonella by acid disinfection of chicks feed. *Veterinary Record* 1985; 116: 502.

Hrangkhawl T, Mukhopadhayay SK, Ganguly S, Niyogi D. Effect of growth promoters on broiler birds under experimental supplementation in feed. J *Chem Biol Physical Sci* 2013;Section-B [Biological Sciences] 3(3): 1875-79.

Iji PA, Tivey DR. Natural and Synthetic oligosachharides in brolier chicken diets, *World's Poultry Science Journal* 1998; 54: 129-43.

Izat AL, Adams MH, Cabel MC, Colberg M, Reiber MA, Skinner JT, Waldroup PW. Effects of formic acid or calcium formate in feed on performance and microbiological characteristics of broilers. *Poultry Science* 1990; 69: 1876-82.

Kaniawati S, Skinner JT, Waldroup PW, Izat AL, Colberg M. Effects of feeding organic acids to broilers on performance and salmonella Colonization of the caeca and or contamination of the carcass. *Poultry Science* 1992; 71 (Suppl.1): 159.

Liu JR, Lai SF, You B. Evaluation of an Lactobacillus reuteri strain expressing rumen fungal xylanase as a probiotic for broiler chickens fed on wheat based diet. *British Poultry Science* 2007; 48(4): 507-14.

Loddi MM. Probioticos and prebioticos e acidificanate organico. em dietas para frangos de corte [tese]. *Taboticabal* 2003; FCAV, UNE SP.

Loddi MM, Moraes VM, Nakaghi LSO, Tucci FM, Hannas MI, Ariki JA. Mannanoligosaccharide and organic acids on performance and intestinal morphometric characteristics of broiler chickens. In: *Proceeding of the 20th Annual Symposium*. 2004; Suppl.1, 45.

Lon R, Dietary MOS as an approach for altering prevalence of antibiotic resistance and distribution of tetracycline resistant determinants: in fecal bacteria from swine. MS Thesis 1995; University of Kentucky.

Mairoka A, Santin AME, Borges SA, Opalinski M, Silva AVF. Evaluation of a mix of fumaric, lactic, citric and ascorbic acids on starter diets of broiler. *Archives Veterinary Sciences*. 2004; 9(1): 31 – 37.

Mannio PF. Enzyme supplementation of barley based diets for broiler chickens. *Aust. J. Agric. Animal Husbandry* 1981; 21, 296-302.

Mathlouthi N, Junin H, Larbier M. Effects of xylanase and beta-glucanase supplementation of wheat or wheat and barley based diets on the performance of male turkeys. *Brit. Poult Sci* 2003; 44(2): 291-98.

Newman K. Mannan oligosaccharides. Natural Polymers with significant impact on the G.I. microflora and the immune system. In: *Biotechnology in the feed industry proceeding of Alltech's 10ᵗʰ Annual symposium*. 1994; Nottingham University Press, Nottingham, U.K. P.167-74.

Nursey I. Control of Salmonella. *Kraftfutter*. 1997; 10: 415-22.

Parks CW, Grimes JL, Ferket PR, Fairchild AS. The effect of mannanoligosaccharide, bambermycin and virginiamycin on performance of large white male turkey toms. *Poultry Science* 2001; 80, 718-23.

Paul I, Isore DP, Joardar SN, Roy B, Aich R, Ganguly S. Effect of dietary yeast cell wall preparation on innate immune response in broiler chickens. *Indian J Anim Sci* 2013a; 83(3): 307-09.

Paul AK, Mukhopadhayay SK, Niyogi D, Ganguly S. Effect of supplementation of different combinations of organic acids as replacer of growth promoting antibiotic in duck. *IIOAB Journal* 2013b;4(2): 40-44.

Pelicano ERL, Souza PA, Souza HBA, Figueiredo DF, Boiago MM, Carvalho SR, Bordon VF. Revista Brasilaria de Cieneia Avicola. 2005; 7(4): campinas Oct/Dec.

Podolsky DK. Regulation of intestinal epithelial proliferation: a few answers, many questions. *American Journal of Physiologic* 1993; 264: 179-86.·

Radecki SY, Yokoyama MT. Intestinal bacteria and their influence on swine nutrition. In: Miller, E.R., Duane, E.U., Lewis, A.J. Swine Nutrition Boston: Butterworth – Heinemann 1991; pp. 439-47.

Roy HS, Mukhopadhayay SK, Niyogi D, Choudhary PK, Ganguly S. Organic acids as a replacer of growth promoter antibiotics in broilers : Pathological and bacteriological studies on intestine. *Indian J Vet Pathol* 2012; 36(1), 114-16.

Santin E, Mairoka A, Macari M, Grecco M, Sanchez JC, Okada TM, Myasaka AM. Performance and intestinal mucosa development of broiler chickens fed diets containing *Saccharomyces cerevisiae* cell wall. *Journal Applied Poultry Research* 2001; 10: 236-44.

Savage TF, Zakrzewska EI, Andreasen JR. The effects of feeding mannanoligosaccharide supplemented diets to Poultry on performance and the morphology of small intestine. *Poultry Science* 1997; 76(Suppl. 1): 139.

Silverslides FG, Bedford MR. Effects of pelleting temperature on the recovery and efficiency of a xylanase enzyme in wheat based diet. *Poultry Science* 1999; 78: 1184-90.

Sims MD, Dawson KA, Newman KE, Spring P, Hooge DM. Effect of dietary Mannan oligosaccharide, Bacitracin methylene disalicylate or both on the liver performance and intestinal microbiology of Turkeys. *Poultry Science* 2004; 83: 1148-54.

Spring P, Wenk C, Dawson KA, Newman KE. Effect of mannan oligosaccharide on different caecal parameters and on caecal concentration on enteric bacteria in challenged broiler chick. *Poultry Science* 2000; 79: 205-11.

Stanley VG, Gray C, Daley M, Kruegar WF, Sefton AE. An alternative to antibiotic based drugs in feed for enhancing performance of broilers grown on Eimeria sp infected litter. *Poultry Science* 2004; 83(1): 39-45.

Thompson JL, Hinton M. Antibacterial activity of formic and propionic acids in the diet of hens on Salmonella in the crop. *British Poultry Science* 1997; 38, 59-65.

Valdman A, Vahl HA. Xylanase in broiler diets with differences in characteristics and content of wheat. *British Poultry Science* 1994; 35(4), 537-50.

Veeramani P, Selvan ST, Viswanathan K. Effect of acidic and alkaline drinking water on body weight gain and feed efficiency in commercial broiler. *Indian Journal of Poultry Science* 2003; 38: 42-44.

Versteegh HAJ, Jongbloed AW. The effect of supplementary lactic acid in diets on the performance of broilers. Institute for Animal Science and Health. Branch Runderweg, Lelystad, The Netherlands.1999; ID-DLO Rep.No. 99.006.

Visek, W.J. The mode of growth promotion by antibiotics. *Journal of Animal Science* 1978; 46, 1447-69.

Wu YB, Ravindran V, Thomas DG, Birtles MJ, Hendriks WH. Influence of Phytase and Xylanase, individually or in combination, on performance, apparent metabolisable energy (AME), digestive tract measurements and gut morphology in broilers fed wheat based diet containing adequate level of phosphorus. *British Poultry Science* 2004; 45: 76-84.

Yubo WB, Changhua L, Shiyan Q, Limin G, Wenqing L, Defali, Properties of *Aspergillus xylanase* and the effects of xylanase supplementation in wheat based diets on growth performance and the blood biochemical values in broilers. *Asia-Australian Journal of Animal Science* 2005; 18: 66-74.

Aguilar-Palazuelos E, de J. Zazueta-Morales J, Martínez-Bustos F. 2006. Preparation of high-quality protein-based extruded pellets expanded by microwave oven. *Cereal Chemistry,* 83(4) 363–69.

Conway HF. 1971. *Extrusion* cooking of cereals and soybeans. *Food Product Development,* 5(2) 14–17.

Della Valle G, Vergnes B, Colonna P, Patria A. 1997. Relations between rheological properties of molten starches and their expansion behaviour in extrusion. *Journal of Food Engineering,* Vol.31 pp.277–96.

Kokini JL, Ho C-T, Karwe MV. 1992. *Food Extrusion Science and Technology.* New York. Marcel Dekker.

McPherson AE, Bailey TB. Jane J. 2000. Extrusion of cross-linked hydroxypropylated corn starches I. Pasting properties. *Cereal Chemistry,* 77, 320-25.

Parsons MH, Hsieh F, Huff HE. 1996. Extrusion cooking of corn meal with sodium bicarbonate and sodium aluminum phosphate. *Journal of Food Processing and Preservation,* 20(3) 221-34. DOI: 10.1111/j.1745-4549.1996.tb00744.x